AGRICULTURAL WATER PRODUCTIVITY OPTIMIZATION FOR IRRIGATED
TEFF *(Eragrostic Tef)* IN A WATER SCARCE SEMI-ARID REGION OF ETHIOPIA

Yenesew Mengiste Yihun

Thesis committee

Promotor

Em. Prof. Dr E. Schultz
Emeritus Professor of Land and Water Development
UNESCO-IHE Institute for Water Education
Delft, the Netherlands

Co-promotors

Dr Teklu Erkossa Jijo
International Researcher, Land and Water Resources
International Water Management Institute (IWMI), East Africa and Nile Basin Offices
Addis Ababa, Ethiopia

Dr Abraham Mehari Haile
Senior Lecturer in Hydraulic Engineering
UNESCO-IHE Institute for Water Education
Delft, the Netherlands

Other members

Prof. Dr P.J.G.J. Hellegers, Wageningen University
Prof. Dr N.C. van de Giesen, Delft University of Technology
Dr Eyasu Yazew Hagos, Mekelle University, Ethiopia
Dr F.W.M. van Steenbergen, MetaMeta Research, 's Hertogenbosch

This research was conducted under the auspices of the SENSE Research School for Socio-Economic and Natural Sciences of the Environment.

AGRICULTURAL WATER PRODUCTIVITY OPTIMIZATION FOR IRRIGATED TEFF (*Eragrostic Tef*) IN A WATER SCARCE SEMI-ARID REGION OF ETHIOPIA

Thesis

Submitted in fulfilment of the requirements of
the Academic Board of Wageningen University and
the Academic Board of the UNESCO-IHE Institute for Water Education
for the degree of doctor
to be defended in public
on 22 January, 2015 at 11.00 a.m.
in Delft, the Netherlands

by

Yenesew Mengiste Yihun
born in Addis Ababa, Ethiopia

CRC Press
Taylor & Francis Group
Boca Raton London New York

CRC Press is an imprint of the
Taylor & Francis Group, an **informa** business

A BALKEMA BOOK

First issued in hardback 2018

CRC Press/Balkema is an imprint of the Taylor & Francis Group, an informa business

© 2015, Yenesew Mengiste Yihun

Published by:
CRC Press/Balkema
PO Box 11320, 2301 EH Leiden, The Netherlands
e-mail: Pub.NL@taylorandfrancis.com
www.crcpress.com – www.taylorandfrancis.com

ISBN 13: 978-1-138-37328-0 (hbk)
ISBN 13: 978-1-138-02766-4 (pbk)

Table of Contents

List of Figures

List of Tables

Acknowledgements

I am extremely grateful to my supervisor Em.Prof.dr.ir. E. Schultz for his wise guidance, critical and innovative insights, wealth of broad knowledge and understanding, and very strong commitment towards this study that played a pivotal role in the success of this endeavour. I have learned a lot from him professionally as well as personally, which has significantly improved my professional capabilities and greatly enriched my life personally. He has been the best professor I can talk of. His encouragement, advice, corrections and concern for my welfare and health are very much appreciated. I shared inspiring discussions during his field supervision visit to Ethiopia which made me to develop tremendous confidence in my work. I never forget the true story he told me about women can do best with their top quality of commitment.

I would like to extend my sincere thanks to my co-supervisor Dr. Abraham Mehari Haile, MSc for all the help he gave me and the guidance he kindly extended to me during this study. I benefited a lot and owe many thanks for his significant contribution to this research and his quest for innovative water management applications, his dynamic and humble personality, kind and gentle behaviour, ability for critical thinking, technical writing ability, and for appealing always positive, friendly and helpful, which are the most virtues that will remain exemplary and worth following in the future.

Many thanks are due to my co-supervisor from the International Water Management Institute (IWMI). Sincere gratitude is due to Dr. Teklu Erkossa Jijo, who happens to be an expert in my study area. I exceedingly benefited from his professional valuable insights, smoothness all the ways and creating a comfortable environment while I was in Ethiopia for my experimental work and data collection in Melkassa Agricultural Research Centre, as well as for providing space to work in the office of IWMI, all of which greatly helped in the successful completion of this study. I owe many thanks for his consistent encouragement to pursue excellence in every component of this study.

During this study, I worked at UNESCO-IHE, Delft, the Netherlands, International Water Management Institute (IWMI), Addis Ababa and Melkassa Agricultural Research Centre, Nazareth, Ethiopia. I would like to extend my sincere thanks to the staff members of IWMI, specifically of W/O Nigist Wagaye, programme management officer and W/O Yenenesh Abebe, GIS Department, for providing the Aridity Index map of Ethiopia and the staff members of Melkassa Agricultural Research Centre for providing their good office and all the necessary equipment to help my field research to be more scientific and interesting. My sincere thanks go to Dr. Tena Alamerew, Ato Worku Gizaw (Addis Ababa University) and Dr. Ketema Tilahun (Charles State University) for their unlimited support and availability in my hard times. I am very grateful to Dr. Tilahun Hordoffa, for his professional advice and practical support. I would like to extend my thanks to Daniel Bekele, Gezachew Legesse and Mezegebu (PhD student in Wageningen University) for showing interest in my work and for the valuable discussions, intellectual and technical support and to ease all the difficulties I had at the time of experimental field data collection. I owe great respect and thanks to my field assistants Ato Yelma Abebe and Ato Mitiku. Their commitment and cooperation is highly appreciated.

I would like to extend my heartfelt thanks to the ever available support and kind interaction to many of the UNESCO-IHE colleagues. Special thanks to Jolanda Boots, Tonneke Morgenstond, Marielle van Erven, Maria Laura Sorrentino, Sylvia van Opdorp-Stijlen, Jos Bult, Anique Karsten and Peter Heerings for their very kind cooperation in bad and good days. I feel great honour to thank all Ethiopian friends in UNESCO-IHE who have made my life an enjoyable experience during my stay in the Netherlands. Special thanks go to Ermias Tefere, Dr. Girma Yimer, Adey Nigatu, Yared Ashenafi, Fikadu, Seleshi Getahun, Dr. Solomon, Rahel, Abonesh and Eskinder. All the understanding and encouragement they had directly or indirectly contributed to the successful completion of the study.

The success of this dissertation would not have been possible without the financial support from the Netherlands Organization for International Cooperation in Higher Education (NUFFIC), which is gratefully acknowledged. Moreover I would like to extend my gratitude to the International Foundation for Science (IFS) for providing additional funds for the field research.

Finally, I would especially like to express deep appreciation to my loving and caring parents, my mother Yanchiamlak Ferede and my father Mengiste Yihun, and to all my sisters (Etaye, Yeme, Mafi and Richo) and brothers (Bereket (Molx), Ellu, Davis, Baby and Besu) for their unconditional love, unbelievable support and prayers. In my early years they have always guided me softly in the right directions and they have always been there for me even after I gave birth to a beautiful boy. I never forget the moment they told me while I hesitated to go abroad to start my study leaving my son on those early ages (3 months). Special thanks go to my elder brother, Bereket Mengiste to take care of my son and to help my mother and all the family in my absence. He always created the environment to continue expanding my knowledge.

I am greatly indebted to my husband Shimelis Berhanu for his deep love, continuous caring, understanding, and for the very long endurance, which provided me energy to successfully complete this challenging venture. I wish to recognize his supportive feedback and intellectual suggestions on solving many problems. I would also like to thank our lovely son to stay calm and comfortable with my mom.

Above all I thank everyone who contributed from the beginning to successful completion of this thesis.

This thesis is dedicated to my mom, Yanchiamlak, my brother Bereket, my husband Shimelis and our son Johannes Shimelis.

Yenesew Mengiste Yihun
Delft, the Netherlands

Summary

The world population is expected to grow from 7 billion at present to 9 billion by 2050. The standard of living in the emerging countries (almost 75% of the world population) is rapidly rising. In the least developed countries there is generally a rapid growth of the population. The population growth combined with the rise of living standards requires a substantial increase of food production to ensure sustainable food security.

Today's agriculture sector faces in fact a complex series of challenges: to produce more food of better quality while using less water per unit of output, to produce sustainable diets and to reduce malnutrition, to apply clean technologies that ensure environmental sustainability, to cope with possible impacts of climate change and to contribute in a productive way to the local and national economy, from producer to consumer. Thus improving water management practices in agriculture is crucial for maintaining global food security and alleviating rural poverty. Irrigated land now accounts for about 20% of the world's farmed area and takes care for 55% of global food production. Globally, irrigation represents 70% of diverted water. Reducing irrigation water use by increasing irrigation water productivity is crucial to meet the ever increasing water demand for agriculture. Thus, irrigated agriculture is faced with pressures to decrease its share of water usage, while at the same time producing sufficient food and fibre for a growing population and other needs.

The rapid increase in population necessitates adequate management of Ethiopia's land and water resources. The population of Ethiopia has increased from 24 million in 1970 to 85 million in 2012 with population growth rate of 3.2% and an expected population of 145 million by 2050. The distribution of the Ethiopian population is generally related to the agro-ecological characteristics and favourable topographic conditions of the area. The economy of the country is highly dependent on agriculture, which is in turn dependent on the availability of seasonal rainfall. Agriculture is unthinkable without land and water. Agriculture can be effective only when it gets sufficient water at the right time. About 85% of the Ethiopian population is dependent on rainfed agriculture. Inadequate seasonal rainfall can cause serious food shortage that can destabilize the social and economic life of the people. Most regions of Ethiopia are suffering from insufficient and unreliable rainfall. Due to severe soil erosion in the highlands, soil of this region is shallow in depth, there is a very low moisture retention capacity and low organic matter content. These adversities of climate and soil result in the prevalence of soil moisture deficit for most of the year, which leads to the loss of crop production. In most regions of Ethiopia rain occurs only in few months of the growing season and is usually short and intense resulting in high runoff, which goes unutilized. The intensity of recurrent droughts affects the livelihoods of agricultural communities and the whole economy. The situation is exacerbated by the declining quality of water and soil. Thus food insecurity has remained the major problem that is a great concern to the country. Therefore it is imperative to bring large areas of the arid, semi-arid and sub-humid regions with uneven rainfall distribution under irrigation.

Irrigation has a multi-faceted role in contributing towards food security, self-sufficiency, food production and export. Scarce water resources and growing competition for water will reduce its availability for irrigation. At the same time, the need to meet the growing demand for food will require increased crop production from less water. Achieving greater efficiency of water use will be a primary challenge in the near future and will include the application of techniques and practices that deliver a more accurate supply of water to crops.

In addition, as the population density in the highlands of Ethiopia continued to increase, more and more marginal lands were put under cultivation, which eventually resulted in severe degradation of the agro-ecological resource base and declining agricultural production. Consequently, population expansion increased towards the extensive lowland (arid and semi-arid) regions. Unfortunately, regions where water is scarce face uncertainties in water supply due to periodic droughts that may severely impact water resources and threaten their sustainability for optimum production. Increasing productivity of water in agriculture by producing more output under more efficient use of water is a key strategy for addressing water scarcity. In addition increase in irrigated area, cropping intensity and crop yields have helped to stabilize food production per capita, even though population and per capita food intake have grown significantly. This calls for the use of suitable innovative technologies for improved and sustainable agricultural production and productivity.

As water supplies are limited, the farmer's goal is to maximize net income per unit of water used. Since there is a reasonable increase in the price of agricultural production, great attention is required to increase water productivity (WP). Recently, emphasis has been placed on the concept of water productivity, defined here as the productivity of water in a given use in terms of quantity and quality of water diverted. Given this, the term water productivity refers to the amount or the value of product over volume or value of water diverted. Depending on how the terms in the numerator and denominator are expressed, water productivity can be expressed in general physical and economic terms. Physical productivity is defined as the quantity of the product divided by the amount of water diverted (kg/m^3). Economic water productivity is defined as value per unit of water or the net present value of the amount of the product divided by the net present value of the amount of water diverted. Increasing the productivity of water in agriculture will play a vital role in easing the competition for scarce water resources, prevention of environmental degradation and provision of food security. Molden and Rijsberman give a simple argument to the above statement: by growing more food with less water, more water will be available for other natural and human uses. It is thus imperative to examine the existing water allocation and distribution systems as well as cropping patterns and intensities, in order to identify the actual constraints that require improvement. Hence, this research was aimed at developing a strategy to use the limited amount of available water as efficiently as possible and to help identify innovative technological water management approaches and practices for optimum agricultural production and water productivity.

A large number of crops are grown in Ethiopia. That includes cereals (Teff, Wheat, Barley, Corn, Sorghum and Millet), pulses, oilseeds, vegetables, root and tubers, fruits, fibers, stimulants and Sugarcane. From the cereals, Teff is known to be a rainfed crop and produced only once in a year. As it is only produced once in a year, the supply didn't much with its demand and the current price increased five times compared with the price in 2006. In 2010, the Teff price was still above US$ 650/ton, creating hardships for many Ethiopian families, who were forced to switch to other cereals as substitutes. Still, Teff has remained the preferred food, as evidenced by the persistently high prices over the past five years. As it is very important throughout the country, a number of biological genetic intensification ways was used to increase the genetic makeup of Teff and to increase the supply of the crop in order to maintain the price. Within the past 20 years, Teff productivity has increased by about 25-30%, with three-fourths of this gain attributed to the introduction of the few improved varieties that have been released and the remainder was by improved agronomic practices, mainly the application of 64 kg of di-ammonium phosphate (DAP) and 46 kg/ha of Nitrogen in the form of urea. Currently, in the experimental plots, Teff yields have reached to 3.3 ton/ha. Maximum yields obtained by farmers in the farmers field is 2.5 ton/ha.

However, typical farmer yields are around 1 ton/ha of grain and 5-6 ton/ha of straw used as animal feed.

Teff (*Eragrostic Tef*) is the major indigenous cereal crop of Ethiopia, where it originated and was diversified. It is a highly demanded cereal and a staple food grain for more than 85% of the 85 million people of Ethiopia. Teff has higher market prices than the other cereals for both its grain and straw. The farmers earn more for growing Teff than for growing other cereals. Farmers and commercial growers produce Teff for local and export markets. More than half of the area under cereals is for Teff production. Teff grain is not attacked by weevils, which means that it has a reduced post harvest loss in storage and requires no pest-controlling storage chemicals. Many investigations have been carried out and indicate that Teff is adapted to environments ranging from drought-stressed to waterlogged soil conditions.

Teff flour is primarily used to make a fermented, sour dough type, flat bread called Injera. Teff is also eaten as porridge or used as an ingredient of home-brewed alcoholic drinks. It has a high iron content and high potential as export crop to USA and European countries as it contains no gluten and is considered a healthy food grain. Approximately one million Americans suffer from celiac disease (gluten sensitivity) and Teff may provide a niche for meeting the dietary requirement. Serious attempts are underway to expand its cultivation in Europe, notably in the Netherlands and in the USA.

For centuries and until 5 years ago, Teff has only been grown as a rainfed crop and produced only once in a year a maximum yield of 1 ton/ha. As the price of Teff increased over the last decade from 2,000 Birr (about US$ 130) per ton to 10,000 Birr (US$ 650) per ton, increasing the production of the crop including using irrigation has become important.

To move away from exclusively rain dependent agriculture is a way to combat frequent crop failure. Two types of field experiments were done to see the effect of irrigation on Teff production. The field experiments were carried out in the dry seasons of 2010/2011 and 2011/2012. Different irrigation management and crop sowing rates were used as management scenarios to determine the optimum amount of irrigation water applied and the sowing rate needed to grow Teff. The sensitive stages of the growth season to the imposed water stress were identified. In addition, also at Melkassa Agricultural Research Centre, two years experiments with four lysimeters were conducted with irrigated Teff during the dry seasons.

The detailed investigation to examine the response of Teff yield, biomass and crop water productivity value for the different irrigation water application scenarios was conducted as follows. A selected combination of depth of irrigation water application (amount) and growth stage (time) of Teff (*Eragrostic Tef*) was used as experimental design in order to determine the optimum water application depth at specific growth stages that result in optimum crop water productivity (CWP). This research investigated the sensitivity of each growing stage to drought stress in detail. Four different levels of irrigation water supply were applied, full crop water requirement 0% deficit (ETc), 25% deficit (applying 75% of crop water requirement), 50% deficit (applying 50% of crop water requirement) and 75% deficit (applying 25% crop water requirement).The phenological cycle was divided into phases which are considered to be most relevant from the viewpoint of their response to irrigation, i.e. initial stage (P1), development stage (P2), mid season stage (P3) and late season stage (P4). A four by four combination of sixteen treatments with three replications was set in the experimental field to make a total of forty-eight trials. Each set of these 48 trials was tested at seeding rates of 10 and 25 kg/ha. Thus the total field experimental plots established in Melkassa Agricultural Research Centre for the first experimental season were 96. Each experimental field had an area of 8 m^2 (2.0 x 4.0 m^2). The individual fields were separated from each other by means of soil bunds.

Teff cultivar, locally called, *Kuncho* was selected and its CWP was assessed under the 16 different treatments. *Kuncho* was selected because as compared to the other local varieties released by Melkassa Agricultural Research Centre, it is the most favourite among the locals and has high market value within the country and the region. The assessments were done based on actual grain and biomass yields obtained during two irrigation seasons: 1) November 2010 to March 2011; and 2) December 2011 to April, 2012. The dates of the main phenological stages, like sowing date, date of 90% emergence, 50% flowering, duration of flowering, senescence and maturity were recorded. Plant height was measured in every ten days interval from the fixed sample of each experimental plot. Above ground biomass observations were also made in every ten days from undisturbed sample areas of 1 m^2. The above ground biomass was dried with an oven drier for 48 hours at 60 ^{0}C and then weighted. Grain yields were measured after maturity from pooled samples of an area of 2.0 x 3.0 m^2 in each plot. The crop was harvested manually. The grain and total biomass fresh weight were weighted at maturity and then dried and weighted on a sensitive balance. Soil moisture was regularly monitored using a Neutron Probe. Irrigation was applied in accordance with the different water deficit conditions. In the case of a full irrigation application (no deficit), the soil moisture was constantly kept at field capacity. The Neutron Probe was calibrated using the Gravimetric Method whereby soil samples were collected from soil depths of 15, 30 and 45 cm before sowing and in every 5-7 days interval up to maturity. The measured soil moisture in weight basis was multiplied by the bulk density to convert to volume basis.

It was found that decreasing the generally recommended seeding rate of 25 kg/ha to 10 kg/ha increased the strength of the Teff stalk and resulted in a greater resistance of the stem against lodging because of heavy panicles at maturity. In addition the smaller seeding rate increased the tillering potential. For the seeding rate of 10 kg/ha, a maximum yield of 2.91 and 3.05 ton/ha respectively was obtained from the treatment which received the optimum crop water requirement during the experimental seasons. Whereas, grain yields of 3.12 and 3.3 ton/ha were obtained for the seeding rate of 25 kg/ha. This is three fold the yield farmers currently harvest from rainfed cultivation. Generally, regardless of the seasons, the patterns of response to irrigation treatment were similar and showed a significant and positive response. During 2010/2011, when 100% of ETc was applied (0% deficit - treatment T1), grain yield of Teff for the seeding rate of 25 kg/ha and 10 kg/ha was found to be 3.12 and 2.91 ton/ha respectively. In case of treatment T2 (75% of ETc irrigation application i.e. 25% deficit) the yield values were reduced to 2.45 and 2.27 ton/ha respectively. Much more significantly lower yields of 0.69 and 0.45 ton/ha were obtained for treatments with 75% irrigation deficit treatment (T4) throughout the whole growth stage for both 25 kg/ha and 10 kg/ha seeding rates. Moreover, 75%, 50% and 25% irrigation water reduction throughout the whole growth stage decreased the Teff yield by 77.9%, 51.6% and 21.5% respectively. Treatments T7 (75% deficit) III, T11 (50% deficit) III and T15 (25% deficit) III which were conducted under adequate watering conditions throughout the first two periods of the growing season, and followed by a period of stress at the mid season stage with 75%, 50% and 25% deficit water application resulted in the second, the third, and the fourth lowest yield respectively by 30%, 23% and 23% for the seeding rate of 25 kg/ha. This yield reduction is significant compared with stressing the crop during late season stage having a reduction of 11%. Stressing the crop either by one-half or three-quarters at the mid season stage, resulted in lower yields next to stressing the crop throughout the growth season. A maximum water deficit of 50% during the late season stage had an insignificant impact on Teff yield and water productivity. As the price of Teff seed increased, decreasing the seeding rate can save seed. Higher crop water productivity values were obtained with the treatment receiving 75% ETc. The obtained values varied with the years and ranged from 1.12 to 1.16 kg/m^3 and 1.08 to 1.31 kg/m^3 for the seeding rate of 25 kg/ha and 10 kg/ha respectively. The yield and crop water productivity differences were insignificant between full irrigation and 25% deficit irrigation distributed throughout the growth period. Thus, when water is scarce and irrigable land is relatively abundant, as is the case in Ethiopia, adopting the 25% water deficit irrigation with 10 kg/ha seeding rate is recommended.

Despite of the importance, Teff as a food crop in Ethiopia, there has been only limited research on its agronomic and physiological responses to water and other physical stress. Moreover, in arid and semi-arid regions such as the Central Awash Rift Valley (the focal study area), as water is scarce increasing CWP is imperative for sustainable food and water security. However, accurate determination of the crop coefficient (Kc), the ratio between crop evapotranspiration (ETc) and reference evapotranspiration (ETo) is a key factor for effective irrigation planning and management.

Crop coefficient (Kc) for Teff was accurately determined using the lysimeters. The lysimeters were of the non-weighing type, two of them were square in cross-section with 2 m length, 2 m width and 2 m depth and the other two were rectangular with 2 m length, 1 m width and 2 m depth. They were constructed of reinforced concrete and the inside was lined with a plastic sheet to avoid leakage or lateral inflow and outflow of water. The stored soil moisture, which is one of the important inputs for the calculation of crop evapotranspiration (ETc), was monitored with a Neutron Probe on every alternate day based on the root depth of Teff. For greater accuracy, the Neutron Probe was calibrated for the soil type of the experimental plots using the Gravimetric Method by establishing wet and dry points, to obtain a wide range of moisture and to make it possible for the probe to read the ranges. Irrigation was applied to the crop when about 55% of the available soil moisture was depleted from the effective root zone, as recommended by Allen et al. (1998). The depletion was converted into volume basis, which was manually supplied using a watering can with a known volume.

The actual Teff ETc was measured using the water balance of the four lysimeters, the overall average ETc was found to be 320 mm for the whole growth stage of the crop. The ETo was calculated using the daily weather data of the study area by using software called ETo calculator. The obtained Kc values were 0.6, 0.8, 1.1 and 0.8 for the initial, crop development, mid season and late season stages respectively. Compared to the Kc values of Barley and Millet, known and used as a substitute for Teff, the Kc values obtained for Barley and Millet for the different growth stages were respectively recorded as follows 0.3, 0.75, 1.05-1.2 and 0.25-0.4. The large difference in the values of Kc in the initial and late season stage demonstrates the importance of regional determination of the crop itself, other than using the Kc value of similar crops.

Elaborating irrigation schedules merely on the basis of field research is rather difficult and time consuming, application of models is required. Several crop models are available to simulate yield response to water. However, they are used mostly by scientists, graduate students and advanced users as these models require an extended number of variables and input parameters not easily available for the diverse range of crops and sites. Moreover, these models present substantial complexity and are rarely used by the majority of target users such as extension personnel, water users associations, consulting engineers, irrigation and farm managers, planners and economists.

AquaCrop is a water-driven model developed by the Food and Agriculture Organization of the United Nations (FAO) for use as a decision support tool in planning and scenario analysis in different seasons and locations. Although the model is relatively simple and requires fewer data input than most of the other models, it emphasizes on the fundamental process involved in crop water productivity and in the response to water deficits, both from physiological and agronomic perspective. It is designed to balance simplicity, accuracy and robustness, and is suited to address conditions where water is a key limiting factor in crop production. The key parameters are normalized water productivity, harvest index, canopy cover, yield and biomass. The model can simulate the variation in crop yield and biomass for the different irrigation water scenarios. AquaCrop model has for example been applied to evaluate the effect of changes in the quantity of irrigation water for Quinoa, Wheat, Sunflower and Maize, respectively. Those researchers revealed that it is a model for scenario analysis that provides a good balance between robustness and output precision.

In the present research AquaCrop was calibrated for Teff in the dry seasons of 2010 and 2011. Independent data sets were used for validation of the model. Model validation was undertaken for the different water stress levels. For treatments receiving less amount of water stress, the model confirmed that there is a good agreement between simulations and observations with coefficient of determination $(r^2) = 0.80$, index of agreement $(d) = 0.94$ and root mean square error (RMSE) = 13.9. As the water stress level increases, the simulated canopy cover, biomass and grain yield were underestimated with $r^2 = 0.39$, $d = 0.45$ and RMSE = 33.6. Moreover, for those treatments receiving higher stress level the observed mean is a better predictor than the model. Model validation revealed that a limited number of inputs are required to model yield response of Teff to soil water availability in the Central Rift Valley. The AquaCrop model balanced between limited parameterization and good accuracy, and it is therefore an effective tool to study different water management scenarios. Therefore, this model can be used to simulate the water management effects on yield and handle managements that increase water productivity.

Overall, the research has demonstrated the potential and the limitations of combining experimental fieldwork with modelling to optimize agricultural water productivity for Teff cultivation. Focusing on only experimental fieldwork is a single approach, and is hardly ever sufficient for achieving the best solutions to the current water management problems. New guidelines on using the combined effort of experimental work in the field to produce field experimental data and using models are clearly needed. It is to these needs as well as to the required increase of Teff production under water scarce conditions that this research has provided its main contribution.

1 Introduction

That fresh water resources are not infinite is clearly demonstrated in river basins where, through increased water withdrawals for the expansion of irrigated agricultural areas, rivers fail to reach the sea, i.e. closed basins. Typical issues in such closed basins are environmental degradation, declining groundwater tables, intrusion of seawater in aquifers and deterioration of the ecological state of wetlands (Molle et al., 2010). Industries are demanding more water, and growing populations require more water for domestic use. The production of food in agricultural systems, whether in rainfed or irrigated areas, takes water from the system that is not available for later reuse. To supply water to agricultural fields for the evaporation (from the surface) and transpiration (from the plant) process, water is diverted from rivers, pumped from groundwater reservoirs, or harvested from the rain. Excess water infiltrates in the soil and returns to the system where it may be available for reuse (Perry, 2007). In the context of a growing population, a changing climate and increasing competition for water, it is unlikely that agriculture can secure a larger share of the fresh water resources.

Agriculture is unthinkable without land and water. Agriculture can be effective only when it gets sufficient water at the right time. In Ethiopia, about 85% of the Ethiopian population is dependent on rainfed agriculture. Most regions of Ethiopia are suffering from insufficient and unreliable rainfall. In most regions rain occurs only in a few months of the growing season and is usually short and intense resulting in high runoff, which goes unutilized. The intensity of recurrent droughts affects the livelihoods of agricultural communities and the whole economy. The situation is exacerbated by the declining quality of water and soil. In the past two decades, however, growing population pressure in the highland areas with rainfed agriculture on rapidly dwindling water resources and a declining natural resource base has placed irrigated agriculture a prominent position on the country's development agenda. The importance of irrigation to enhance agricultural production in the water stressed regions of Ethiopia is yet to be translated into real investment in and actual development of irrigated agriculture. Currently, irrigation accounts for only 3% of the total 10 ha million arable lands and virtually all food crops including Teff (focus of this research) are rainfed. Thus food insecurity has remained the major problem that is a great concern to the country. Therefore it is imperative to bring large areas of the arid, semi-arid and sub-humid regions with uneven rainfall distribution under irrigation. The study area Central Rift Valley is well known for irrigation activities.

An increase in agricultural water productivity is an important component to sustain and improve food production towards ensuring sustainable food security in the food insecure region. Recently, emphasis has been placed on the concept of water productivity, defined here as the productivity of water in a given use in terms of quantity and quality of water diverted. Given this, the term water productivity refers to the amount or the value of product over volume or value of water diverted. Increasing crop water productivity (CWP) as argued by the Food and Agriculture Organisation of the United Nations (FAO) (2010) and Geerts and Raes (2009) can be an important pathway for poverty reduction. This would enable growing more food and hence feeding the ever increasing population of Ethiopia or gaining more benefits with less water thus enhancing the household income. Moreover, more water will be available for other natural and human uses. This strategy is more popularly stated: to produce more crops per drop (Kijne et al., 2003). In a broader sense, increasing the productivity of water means getting more value from each drop of water. Water saving techniques such as water deficit irrigation have in particular been increasingly viewed as one of the best alternative methods of irrigation to intensify the staple food production and water productivity of the vast arable land of Ethiopia. Elaborated strategies developed by crop growth models can be used for designing and better understanding the impact of the aforementioned strategies. CWP can also be improved by proper irrigation scheduling, which is essentially governed by crop evapotranspiration (ETc).

This study aimed at developing a strategy to use the limited amount of available water as efficiently as possible and helps to identify innovative technological water management approaches and practices for optimum agricultural production and water productivity. Experimental field trails and crop water productivity modelling were used.

1.1 Thesis structure

This thesis is presented in eight chapters. The present chapter has introduced the research and presents the structure of the thesis. Brief descriptions of each chapter will be given as follows:

Chapter 2 provides the research background, the general overview of water resource potential from global to national perspective. Moreover, water productivity and crop water productivity - basics investigated with interrelated and cascading sets of definitions useful for different implication on the current water scarcity and food security framework will be discussed.

Chapter 3 general methodology and the clear methodological framework of this research study will be discussed and set in this chapter.

Chapter 4 gives the description and detailed characterization of the study site and will discuss the irrigation development across the Awash River Basin and the whole irrigation systems employed in the river basin.

Chapter 5 focuses on the water productivity and crop water productivity (CWP) of irrigated agriculture in comparison with rainfed agriculture. CWP is imperative for sustainable food and water security. The crop water productivity of Teff (*Eragrostic Tef*), a staple food crop in Ethiopia and an important export crop, with respect to full and limited irrigation conditions at field scale was investigated. The production of yield and biomass of Teff grain in relation to the amount of moisture available under full and drought conditions will also be discussed on this chapter.

Chapter 6 reviews and summarizes several studies that have been conducted over the years to measure crop coefficients for different crops. The need for regional calibration of crop coefficients under given climatic conditions for a realistic estimation of crop water requirement will be explained. The development of the crop coefficient (Kc) and the calculation of crop evapotranspiration for the target crop Teff will be described.

Chapter 7 will elaborate on irrigation schedules, merely on the basis of field research, which is rather difficult and time consuming. This chapter will confirm the importance of using models. Moreover, the performance of the AquaCrop model for Teff under different regimes of irrigation (full and deficit) and different timing of irrigation water application was evaluated for the dry seasons.

Chapter 8 reviews several studies that have been conducted on Teff in Ethiopia. This chapter will present the shift from purely rainfed crop to irrigated crop and will explain the importance of optimum irrigation application for irrigated Teff. Moreover, it presents how the increase of irrigation water to a certain level has increased the yield, the crop water productivity and economic water productivity of Teff.

Chapter 9 summarizes the conclusion of the research based on the various water management strategies presented in this thesis and suggests recommendations for further research.

2 Background and Objectives

2.1 Country profile

Ethiopia is a country located in the horn of Africa, bordered by Eritrea to the North, Djibouti and Somalia to the East, Sudan and South Sudan to the West, and Kenya to the South. It is the most populous landlocked country in the world and the second most populated nation on the African Continent. The population of the country has increased from 24 million in 1970 to 85 million in 2012 with an average growth rate of 3.2% and if the trend continues unabated, it is expected to soar as high as 145 million by 2050. Situated between latitude 3^0 and 15^0 North and longitude 30^0 and 48^0 East. The country has a total land area of 110 million ha (United Nations department of Economics and Social Affairs Population division (UNDP), 2011).

The country is divided into nine ethnically based administrative regional states (Kililoch, Singular Kilil) and three chartered cities (Astedader Akababiwoch, Singlualr-Astedader-Akababi) (Figure 2.1). The nine regions (and three charted cities, marked by asterisks) are: Addis Ababa*, Afar, Amhara, Benishangul Gumuz, Diredawa*, Gambela, Harari*, Oromoiya, Somali, Southern Nations, Nationality and Peoples Region, Tigray and subdivided into sixty-eight zones (Table 2.1). It is further subdivided into 550 woredas and neighbourhoods (kebele) (United State Central Intelligence Authority, 2009).

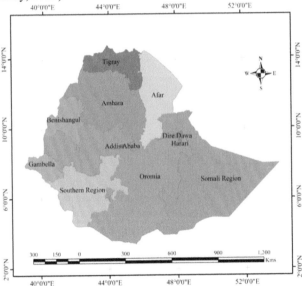

Figure 2.1. Administrative regions of Ethiopia

Ethiopia has a vast highland complex of mountains higher than 4000 m+MSL (mean sea level) including the highest mountain (RasDejen) with 4620 m+MSL and the lowest Dallol (Danakil) Depression (125 m-MSL) in the Afar Regional State. The highland plateaus dissected by the Central Rift Valley, which runs generally Southwest to Northwest and is surrounded by lowlands, steppes or semi desert. The great diversity of terrain determines wide variation in climate, soils, water, natural vegetation and settlement patterns. The variation in topography has created three traditional climatic zones, which have been known since antiquity as '*dega*', '*weina dega*' and '*kolla*'. The dega (also known as a cool zone mostly laying above 2400 m is characterized by daily temperatures ranging from near freezing to 16 ^{0}C and covers the central parts of the western and eastern section of the high plateaus. The weina dega or the temperate zone ranges between 1,500 m and 2,400 m+MSL with a daily temperature ranging between 16 ^{0}C and 30 ^{0}C. The kolla or the hot lowlands generally comprises areas lower than 1500 m in elevation, including the Danakil Depression, and the tropical valleys of the Blue Nile (US Library of Congress, 2005). Potential evapotranspiration varies between 1,700-2,600 mm in arid and semi-arid areas and 1,600-2,100 mm in dry sub-humid areas. Average annual rainfall is 848 mm, varying from about 2,000 mm over some parts of Southwest Ethiopia to less than 100 mm convective storms, and with extreme spatial and temporal variability. This results in a high risk of annual drought and intra-seasonal dry spells.

Table 2.1. Characteristics of the regional states and city administrations of the country (Central Statistical Agency (CSA), 2007).

Region Name	Population	Area (km^2)	Density (km^2)	Capital
Addis Ababa	3,147,000	530	5,936	Addis Ababa
Afar Region	1,389,040	96,707	14	Asaita
Amhara Region	19,120,005	159,174	120	Bahir Dar
Benishangule Gumuz Region	625,000	49,289	13	Asosa
Diredawa	398,000	1,213	328	Diredawa
Gambela Region	247,000	25,802	10	Gambela
Harari Region	196,000	311	630	Harar
Oromiya Region	26,553,000	353,007	75	Adama
Somali Region	4,329,001	279,252	16	Jiiga
Southern Nations, Nationalities, and People's Region	14,901,990	112,343	133	Awasa
Tigray Region	4,334,996	50,079	86	Mekelle

Ethiopia is an ecologically diverse country, ranging from the deserts along the eastern border to the tropical forests in the South. Lake Tana in the North is the source of the Blue Nile. It also has a large number of endemic species, notably the Gelada Baboon, the Walia Ibex and the Ethiopian wolf (or Simien fox). The wide range of altitudes has given the country a variety of ecologically distinct areas; this has helped to encourage the evolution of endemic species in ecological isolation.

2.2 Rainfed and irrigated agriculture

Agriculture produces the food to feed the country's population and fulfils other requirements such as the production of fiber, fuel, ingredients for manufacturing. In Ethiopia, agriculture is the foundation of the country's economy, accounting for half of the gross domestic product (GDP), 83.9% of exports, and 80% of total employment. Ethiopia has great agricultural potential because of its vast areas of fertile land, diverse climate, generally adequate rainfall and large labour pool. Despite this potential, however, Ethiopian agriculture has remained underdeveloped. Because of drought, this has repeatedly affected the country, a poor economic base (low productivity, weak infrastructure, and low level of technology) and overpopulation.

In Ethiopia, agriculture mainly consists of subsistence rainfed crop production, irrigation and pastoralism. Ethiopia relies on low-productivity rainfed agriculture for a majority of its national income; consequently, the importance of the timing and amount of rainfall that occurs cannot be overstated. Three seasons are created because of the seasonal variation and atmospheric pressure system, which are known as Kiremt, Belg and Bega. The Kiremt season is the main rainy season and usually lasts from June to September, covering all of Ethiopia except the southern and the south-eastern parts on which 90% of the total crop production is accounted. The Bega season is the dry season and usually lasts from October to February, during which the entire country is dry and crop production suffers from serious moisture stress (Engida, 2000), with the exception of occasional rainfall that is received in central sections. The Belg season is the light rainy season and usually lasts from March to May. It accounts for 10% of the total crop production in Ethiopia (Central Statics Authority, 2010). The dominant crops grown are mainly cereals, Pulses and oil crops, Enset, Coffee and horticultural crops.

Irrespective of Ethiopia's endowment with potentially huge irrigable land the area of land under irrigation so far is only about 3%, showing that water resources have made little contribution towards the development of the irrigated agriculture sector. It can easily be realized, however, that in addition to the underdeveloped irrigation, the accelerated population growth and the disparity of rainfall distribution the production of sufficient food and food security is almost impossible. On the contrary, a number of studies made in the field confirm that if the country's water resources are developed to cater for irrigation, it would be possible to attain enough agricultural surplus both for domestic consumption as well as for external markets. Export crops such as Coffee, oilseed and pulses are mostly rainfed but industrial crops such as Sugarcane, Cotton and fruit are irrigated. Other irrigated crops include vegetables, fruit trees, Maize, Wheat, Potatoes, Sweet potatoes and Bananas.

2.3 Water resources

2.3.1 Water resource in the world

Water in seas and oceans is unlimited, but fresh water resources are finite and limited. Only 2.5% of the total water volume on earth is fresh water of which nearly 70% is frozen in the icecaps of Antartica and Greenland; most of the remainder is present as soil moisture, or lies in deep underground aquifers as groundwater not accessible to human use. 1% of the world's fresh water (~0.007% of all water on earth) is accessible for direct human uses (Charles, 2002). This is the water found in lakes, rivers, reservoirs and those underground sources that are shallow enough to be tapped at an affordable cost. Only this amount is regularly renewed by rainfall and snow, and is therefore available on a sustainable basis. The water resources are seriously under pressure, even at present, in several regions of the globe. It is anticipated that the situation will be worse during the first half of this century (Cosgrove and Rijsberman, 2000).

The growing world population and developing industries are today involved in a competitive struggle with the agriculture sector, the major water consumer compared to industries and municipalities (International Water Management Institute (IWMI), 2000; FAO, 2002; Rijsberman, 2006). The ongoing process of global warming is thought to contribute to the destabilization of the world's weather systems, affecting people in drought prone regions and in areas susceptible to inundation (Vorosmarty et al., 2000). Contamination not only compromises the water quality and threatens the ecology and biodiversity of the ecosystem; it also lowers the available amount of water (Meinzen-Dick and Rosegrant, 2001).

2.3.2 Water resource in Ethiopia

Ethiopia is rich in water resources. The total annual water resource is estimated at 122 billion m³, of which 76.6 billion m³ drains into the Nile River. The usable groundwater resource is estimated to be 2.6 billion m³/year while very little is developed and exploited. There are also numerous lakes in Ethiopia; the eleven major lakes have a total surface area of about 7000 km². There are twelve river basins (Figure 2.2); 80-90% of Ethiopia's water resources are found in four major river basins (FAO, 2005):

- the Nile Basin (including Abbay or Blue Nile, Baroakobo, Setittekeze/Atbara and Mereb) covers 33% of the country and drains the northern and central parts westwards;
- the Rift Valley (including Awash, Denakil, Omogibe and Central lakes) covers 28% of the country;
- the Shebellijuba Basin (including Wabishebelle and Genaledawa) covers 33% of the country and drains the south-eastern mountains towards Somalia and the Indian Ocean;
- the Northeast Coast Basin (including the Ogaden and Gulf of Aden basins) covers 6% of the country.

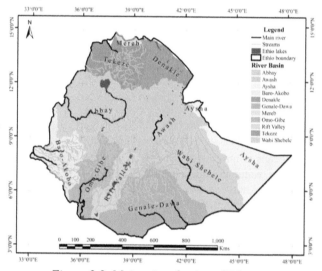

Figure 2.2. Major river basins of Ethiopia

Ethiopia receives an average annual rainfall of 744 mm apparently adequate for food crop production, and pasture growth for livestock. However, the distribution of rain varies from region to region. Much of the eastern parts and countries like Somalia and Djibouti receive very little rainfall, less than 100 mm/year, while the Southwest highlands receive 240 mm/year. In the Southern and Eastern highlands, there is a bi-modal rainfall distribution, with the first and generally smaller rains peaking in April, and the second in September. The main dry season extends from October to February, being longer and drier in the North. Hence, production of sustainable and reliable food crops is almost impossible due to temporal and spatial imbalance in the distribution of rainfall and the consequential non-availability of water in the required period. Intense rainfall sometimes causes flooding, particularly along the Awash River and in the Lower Baroakobo and Wabeshebelle river basins, causing damage to standing crops (Welderufael, 2006). Sometimes, even the Western highlands of the country suffer from food shortage owing to the discrepancies in the rainfall distribution (Ministry of Water Resource (MOWR), 2001; Zhang et al., 2004).

Climate has a direct implication on the development of the river basins with particular influence on the availability and use of water and pattern of settlement. The very high variability exhibited by the climatic components of the country over time and space is the main reason behind the spatial and temporal variability in the availability of water. The shape, size and other physical features of the river basins also contribute to the same. The nature of the river channels, which is a direct function of the relationship between the flow and formation of the channels, governs accessibility to and pattern of use of the waters flowing in the rivers or stored in natural depressions. Table 2.2 indicates important climatic components and water resource potential of basins in Ethiopia in terms of mean annual runoff, storages in major lakes, impoundments and underground.

2.4 Water productivity and crop water productivity – basics

Within the broad definition of water productivity there are interrelated and cascading sets of definitions useful for different purposes. Water productivity is defined as the ratio of the benefits from crop, forestry, fishery, livestock, and mixed agricultural systems to the amount of water required to produce those benefits. In its broader sense it reflects the objectives of producing more food, income, livelihoods and ecological benefits at less social and environmental cost per unit of water used, where water use means either water delivered to a use or depleted by a use (Kijne et al., 2003b). Physical water productivity relates the mass of agricultural output to water used 'more crop per drop'. Economic water productivity relates the economic benefits obtained per unit of water used and has also been applied to relate water use in agriculture to nutrition, jobs, welfare and the environment. Increasing water productivity is particularly appropriate where water is scarce compared with other resources involved in crop production (Schultz et al., 2009). Physical water productivity is presented in detail because it underpins many of the broader concepts and has the largest impact on the amount of water required to produce food.

There are important reasons to improve agricultural water productivity:

- to meet the rising demand for food from a growing, wealthier and urbanized population, in light of water scarcity;
- to respond to pressures to reallocate water from agriculture to cities and to ensure that water is available for environmental uses;
- to contribute to poverty reduction and economic growth. For the rural poor more productive use of water can mean better nutrition for families, more income, productive employment, and greater equity. Targeting high water productivity can reduce investment costs by reducing the amount of water that has to be withdrawn.

Agricultural water productivity optimization in a water scarce semi-arid region of Ethiopia

Table 2.2. Important climatic components and water resource potential of basins in Ethiopia[1]

Basin Name	Temperature (°C)		Rainfall (mm)			Evaporation (mm)	Water resource potential (Bm^3)		
	Min	Max	Min	Max	Average	Average	Surface	Ground	Stored
Wabishebele	6	27	223	1560	425	1500	3	2.3	1.1
Abay	11.4	25.5	800	2220	1420	1300	54	na	30
Genale Dawa	<15	>25	200	1200	528	1450	6	na	-
Awash	20.8	29	160	1600	557	1800	5	0.8	2.2
Tekeze	<10	>22	600	1200	1300	1400	8	na	-
Denakil	5.7	57.3	100	1500	na	Na	0.9	na	na
Ogaden	25	39	200	800	400	Na	0	na	na
Omo-Ghibe	17	29	400	1900	1140	1600	16	1.0	na
Baro-Akobo	<17	>28	600	3000	1420	1800	23	1.0	na
RiftValley	<10	>27	300	1800	na	1610	6	na	56.6
Mereb	18	27	680	2000	na	1500	0.7	0.1	-
Aysha	26	40	120	500	na	Na	0	na	-

Source: Respective Basin Master Plan Studies na: data not available

[1]http://www.mower.gov.et/wresurfacewatertblclimate.php

Water productivity is also sometimes measures specifically for crops (crop water productivity). Irrigated agriculture is the largest water consuming sector and it faces competing demands from other sectors, such as the domestic and the industrial sectors. With an increasing population and less water available for agricultural production, the food securty for future generations is at stake. The agricultural sector faces the challenge to produce more food with less water by increasing crop water prodcutivity. Increasing the productivity of water in agriculture plays a vital role in easing competition for scarce water resources, prevention of environmental degradation and provision of food security (Molden et al., 2003). Molden and Rijsberman (2001) and Rijsberman (2001) give a simple argument to the above statement: by growing more food with less water, more water will be available for other natural and human uses.

The water productivity concept is based on 'more crop per drop' or producing more food from the same water resources' or producing the same amount of food from less water resources'. In broad sense, productivity of water is related to the value derived from the use of water. Definitions of water productivity are not uniform and change with the background of the researcher or stakeholder involved. For example at field scale if we concentrate on the productivity then we can express this as total dry matter production or as actual yield as harvestable product. At basin scale, economists wish to maximize the economical value from water used. Kijne et al. (2003b) provide several starteigies for enhancement of agricultural water productivity by integreting varietal improvement and better resource management at plant level, field level and agro-climatic level. There are several definitions of water productivity, so we have to ask ourselves which crop and which drop are we referring to (Table 2.3).

Table 2.3. Examples of stakeholders and definitions in the water productivity framework (Kijne et al., 2003b)

Stakeholder	Definition	Scale	Target
Farmer	Yield/supply	Field	Maximize income
Agronomist	Yield/evapotranspiration	Field	Sufficient food
Irrigation engineer	Yield/irrigation supply	Irrigation scheme	Proper water allocation
Nutritionist	Calorie/transpiration	Field	Healthy food
Plant physiologist	Dry matter/transpiration	Plant	Utilize light and water resources
Basin policy maker	$/evapotranspiration	River basin	Maximize profits

In this study crop water productivity (CWP) is defined as the amount or the value of product over volume of water depleted or diverted (Molden et al., 2007).

$$CWP = \frac{Y_{act}}{ET_{act}} (kg\ m^{-3})$$ (2.1)

where:
Y_{act} = the actual crop yield (kg/m^3)
ET_{act} = the seasonal crop water consumption by actual crop evapotranspiration (m^3/ha)

The denominator of the water productivity equation is expressed as water either diverted or depleted. Water depleted when it is consumed by evapotranspiration, is incorporated into a product, flows to a location where it cannot be readily reused (to saline groundwater, for example), or becomes heavily polluted (Molden 2003).

When considering this relation from a physical point of view, one should consider transpiration only. The partitioning of evapotranspiration in evaporation and transpiration in field experiments is, however, difficult and therefore not a practical solution. Moreover, evaporation is always a component related to crop specific growth, tillage and water management practices and this water is no longer available for other usage or reuse in the basin. Since evapotranspiration is based on root water uptake, supplies from rainfall, irrigation and capillary rise are integerated.

Water managers tend to be more concerned with the total water input. Rainfed farmers in arid and semi-arid areas, for example, are extremely concerned with capturing and doing the most with limited rainfall. Where an additional supply is available as supplemental irrigation, maximizing the output from a small amount of additional irrigation supply is normally highly productive. For irrigation farmers and managers of irrigation systems, water supply is a managerial factor and they will evaluate their own water productivity (WP) on the basis of canal water supplies in relation to crop yield, rainfall, supplemental irrigation, or full irrigation.

2.5 Yield response factor of crops to deficit irrigation

Water is essential for crop production, and any shortage has an impact on final yields. Therefore, farmers have a tendency to over-irrigate, an approach that runs counter to the conservation of scarce resources. At present, owing to the global expansion of irrigated areas and the limited availability of irrigation water, there is a high pressure to reduce irrigation water use, it is important to optimize the use of water in irrigated agriculture in order to maximize crop yields under frequently occurring situations of deficit irrigation (Garcia-Vila et al., 2009). When water deficit occurs during a specific crop development period, the yield response can vary depending on crop sensitivity at that growth stage (Fereres and Soriano, 2007). Therefore, timing the water deficit appropriately is a tool for scheduling irrigation where a limited supply of water is available. A standard formulation relates four parameters (Y_a, Y_m, ET_a and ET_m) to a fifth: K_y, the yield response factor as follows.

$$K_y = \frac{1 - \dfrac{Y_a}{Y_m}}{1 - \dfrac{ET_a}{ET_m}} \qquad (2.2)$$

where:
Y_a = actual yield (kg/ha)
Y_m = maximum yield (kg/ha)
ET_a = actual evapotranspiration (mm)
ET_m = maximum evapotranspiration (mm)
K_y = yield response factor.

K_y relates relative yield decrease to relative evapotranspiration deficit. Two series of K_y values obtained from FAO data sets and from an International Atomic Energy Agency (IAEA) coordinated research project (CRP) showed a wide range of variation for this parameter $0.20 < K_y < 1.15$ (FAO, 2002), and $0.08 < K_y < 1.75$ (IAEA) (Moutonnet, 2002; Kipkorir et al., 2002).

Many literature reviews relate yield responses of major field crops to deficit irrigation, including Cotton, Maize, Potato, Sugarcane, Peanut, Soybean and Wheat. Crop yields obtained under various levels of reduced evapotranspiration were fitted to linear crop yield response functions (Payero et al., 2009; Abou and Abdrabbo, 2009; Garcia-Vila et al., 2009). Results show that Cotton, Maize, Wheat, Sunflower, Sugarbeet and Potato are well suited to deficit irrigation practices, with reduced evapotranspiration imposed throughout the growing season (Kang et al., 2002; Zhang et al., 2004; Ali et al., 2008). This list may also include Common Bean, Groundnut, Soybean and Sugarcane where reduced evapotranspiration is limited to (a) certain growth stage(s). With a 25% deficit, water use efficiency was 1.2 times that achieved under normal irrigation practices. Irrigation scheduling based on deficit irrigation requires careful evaluation to ensure enhanced efficiency of use of increasingly scarce supplies of irrigation water (Kirda, 2002).

From Equation 2.2 it is possible to calculate Y_a where the available water supply does not meet the full moisture requirements of the crop. Where water deficit occurs during a specific growth stage, the yield response will depend on crop sensitivity during that period. Therefore, the timing of the deficit is a tool for scheduling the use of a limited water supply and in setting priorities among several irrigated crops.

2.6 Problem statement

The food security status of Ethiopia faced a major challenge in the past three decades. Major factors that have contributed to this are continuous soil erosion, nutrient depletion, salinization and other forms of land and water degradation, as well as unreliable rainfall. They also hampered water productivity gain in many areas. Moreover, population pressure and lack of appropriate technological and institutional support are the main driving forces.

Drought has been the main climatic related risk in Ethiopia. Even in good years, Ethiopia cannot meet its large food deficit through rainfed agriculture production. The economic basis of the country is rainfed agriculture. The rainfall is, however, scanty, erratic and inadequate. The topography of the area is undulating. Thus with the traditional agricultural practices, natural resources are severely degraded due to human as well as natural devastation. The level of land productivity is declining at an alarming rate. As a result, the country is not in a position to cover the annual food requirement of the people. In addition farmers in the country lack access to adequate and accurate information and are not linked to markets.

The following points contributed to the low agricultural production and water productivity leading to food insecurity in Ethiopia:

• increasing population and decreasing per capita available arable land and water resources;
• erratic rainfall patterns and inadequate water resource during the dry season, the average productivity of irrigated land is more than ten times that of rainfed production (Makombe et al., 2007);
• poor water management practices in irrigation systems;
• lack of proper agronomic measures;
• inadequacy or lack of drainage facilities in irrigated and rainfed areas.

Innovations that are economically and technologically feasible to small-scale farmers for more effective and rational use of limited supply of water that is available is crucial. Options in irrigation water management and developing new irrigation approaches, not necessarily based on full crop water requirement, but designed to ensure the optimal use of water are important in areas where water scarcity is severe and most of the fertile lands and manpower are idle for long periods outof the major rainy season.

Deficit irrigation practice is one possible irrigation water management strategy that may help farmers to apply limited amounts of water to their crops in time and amount vital for optimum crop water productivity. In order to allocate the scarce water resources among competing uses and to schedule irrigation water, the level of water deficit which maximizes crop water productivity has to be identified. This level of water deficit varies with crop type and cropping pattern, soil type, depth and fertility, climate, water quality as well as type of irrigation system. In addition market accessibility and availability of land and water policy are some of the external factors that may affect crop water productivity.

2.7 Research questions and hypothesis

2.7.1 Research questions

This work attempts to address the following scientific research questions:
- what is the current level of agricultural crop production and productivity under the existing technological and water management practices in rainfed and irrigated agriculture?
- what are the best irrigation technologies and systems as well as water management practices that can significantly increase Teff production and water productivity in Ethiopia from its current level of 0.9 ton/ha?
- which water management and technological practices are best suited to realize optimum agricultural production and water productivity?

2.7.2 Hypothesis

Water productivity, yield and biomass production are functions of irrigation water regimes. As severity of water stress increases, yield and biomass production decreases. Differences exist by varying levels of irrigation water application at various season and such differences are expressed in the yield and biomass component.

Identifying and adopting locally feasible and socially acceptable irrigation technologies and systems coupled with best water management practices can result in a sustainable increase of agricultural water and crop (Teff) productivity under water scarce and deficit conditions.

Given the generally undulating topography, severe soil erosion and arid characteristics of most agricultural areas in Ethiopia, land fertility degradation as well as soil water salinity are real threats to optimize land and water productivity. Current agricultural practices are not tailored to address the problems. Due to these factors, the level of agricultural production and water productivity in Ethiopia is low and it can be significantly improved.

The hypothesis of the study is therefore: improving irrigation water management practices can result in a sustainable increase in agricultural water productivity and production under drought and moisture deficit conditions. More explicitly, water productivity, yield and biomass production are functions of irrigation water regimes. Treatments resulting in more severe stress would have lower yield and biomass production.

To determine whether the hypothesis is supported, the general goals and a series of specific objectives would have to be accomplished as described in the next section.

2.7.3 Research objectives

In humid and sub-humid climatic zones, irrigation is used as a supplemental source of water to optimize crop production. Irrigation is known as a method for maximizing crop production and is insurance for attaining high productivity in the conditions with lack of optimum rainfall to fully meet the water requirement of the crop, which limits the productivity. Due to the growing population and high sectoral competition of water users (agriculture, municipality, industries, recreation, mining, etc.), the quantity of water available to be used in irrigated agriculture is decreasing throughout the country. Moreover, climate change is having an impact in the seasonal distribution and magnitude of rainfall and the recharge of surface and groundwater resources. To ensure sustainability of finite water resources and to optimize agricultural water productivity innovative irrigation water management strategies must be developed and implemented. Therefore, determining proper timing and amount of irrigation is becoming more important for efficient use of water resources and for maximizing crop yield. The overarching scientific objective of this study is therefore to identify innovative and adaptive technologically feasible, socially, and economically acceptable water management approaches and practices for optimum agricultural production and productivity under drought and deficit water supply conditions.

The specific objectives of the study are:

- to determine crop coefficient, crop evapotranspiration, yield and biomass values of Teff (*Eragrostic Tef*);
- to propose, as necessary, technological and water management practices for optimal agricultural water productivity and production;
- to investigate the degree of tolerance of Teff to varying levels of (full and limited) water application at field scale;
- to determine the effect of seeding rates on lodging, yield and biomass production of the crop;
- to evaluate the performance of application of the AquaCrop model for Teff under different irrigation water management practices.

3 Materials and Methods

3.1 Methodological framework

The methodological framework that was followed in this study is schematized in Figure 3.1. The details of the field experimental setup and assessment of water availability in the river basin are specified in this chapter. Elaborating on irrigation schedules merely on the basis of field research is rather difficult and time consuming, justifying the need of using of models is required. A water driven model (AquaCrop) was applied to emphasize the fundamental process involved in crop water productivity and in response to water deficits, both from physiological and agronomic perspectives. They allow studying possible scenarios that might occur in the future to adopt better management strategies. The field experiments and modelling consequently provide a sound scientific basis for planning and the water management scenario analysis. A brief description of the methods used in this study is presented below.

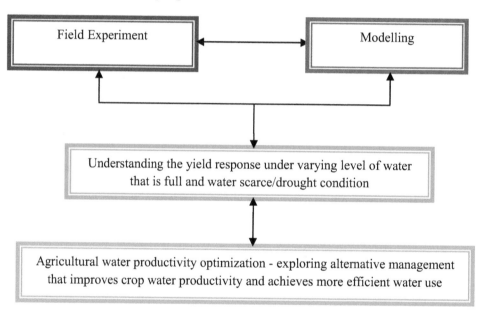

Figure 3.1. Methodological framework followed in this research

3.1.1 Field experiments and assessment

Assessment of current water supply, availability and allocation in the study area

The water supply to the big farm of Melkassa Agricultural Research Centre is Awash River. The quantity of water diverted was quantified by discharge measurements at the intake of the structure by using velocity area methods, using current meters and actual measurements in the intake of the canal by using flumes and partial flumes. The quality of water diverted was tested by direct measurements and through laboratory analyses. In addition salinity hazard and pH have been quantified. The long-term impact of soil and water salinity on crop water productivity has been analyzed using the FAO Five-Point Method (Allen et al., 1998). In addition, primary and secondary data at system level have been collected through field observations, interviews with the help of structured and semi structured questionnaires and focus group discussions with farmers around the research centre to understand how they allocate the diverted water to irrigate their field.

Crop water productivity and Teff yield under different water deficit conditions and seeding rates at experimental field level

The experimental design and setup

The field experiments were carried out in the research farm of Melkassa Agricultural Research Centre in Ethiopia both during the rainy and dry seasons of the years 2010/2011 and 2011/2012. A combination of depth of irrigation water application (amount) and growth stage of Teff (time) was used as experimental design in order to determine the optimum water application depth at specific growth stage that result in optimum crop water productivity and yield. This research investigated the sensitivity of each growing stage to drought stress in detail. Four levels of irrigation water supply including, full crop water requirement (0% deficit (ETc), 25% deficit (applying 75% of ETc), 50% deficit (applying 50% of ETc) and 75% deficit (applying 25% of ETc) were applied at four growing stages of the crop. The crop growing season of Teff (*Eragrostic Tef*) were divided into four major growth periods: initial stage (P1), development stage (P2), mid season stage (P3) and late season stage (P4). Each treatment had three replications making a total of 48 experimental plots that were arranged in a randomized complete block design. The treatments were tested under two seeding rates (10 and 25 kg/ha) with a recommended fertilizer rate. The plot size was 8 m^2 (2.0 m x 4.0 m). The individual fields were separated from each other by means of soil bunds. A watering can with a known volume was used for water application.

Crop coefficient (Kc) and crop evapotranspiration (ETc) determination

All cereal crops have already developed crop coefficient values except Teff crop. FAO recommends two methods for measuring the crop coefficient, single crop coefficient method and dual crop coefficient method (Jensen et al., 1990; Allen et al., 1998). In this study, the crop coefficient was calculated from estimated crop evapotranspiration by the field water balance method using lysimeters, and reference evapotranspiration by using software called ETo calculator, based on the weather data in the area (Raes, 2009). Determining crop coefficient (Kc) requires long-term daily, weekly or monthly weather data of the area for the calculation of reference evapotranspiration (ETo). Lysimeters can directly measure the evapotranspiration rate from the cropped surface. When lysimeters are employed to measure actual evapotranspiration, it is desirable that they should contain an undisturbed representative soil profile. Because in disturbed profile moisture retention and root distribution is likely to be different from that of the original profile and measurements may not be representative. ET can be determined by measuring the various components of the soil water balance. The procedure of this method consists of monitoring the incoming and outgoing water flux into the crop root zone over growing periods (Allen et al., 1998). Lysimeters can also hydrologically isolate soil from surrounding soil and make it possible to eliminate surface and sub-surface flow into and out of the root zone (control volume). By isolating the crop root zone from its environment and controlling the processes that are difficult to measure, the different terms in the soil water balance equation can be determined with greater accuracy as follows. According to Allen et al. (1998) the crop evaporation (ETc) over an interval of time can be found from the water balance equation as:

$$ET_c = I + R - D - S \tag{3.1}$$

where:
ETc   = Evapotranspiration of the crop (mm)
I   = Irrigation (mm)
R   = Rainfall (mm)
D   = Drainage collected (mm)
S   = Decrease in storage of soil moisture (mm)

Based on regular soil moisture records, irrigation is applied when 55 Vol% soil moisture is depleted from crop root zone (Allen et al., 1998). Amount of water to be applied to refill the depleted moisture will be calculated and given to the crop in and outside the lysimeters. Water will be applied in known volume of watering can be converted to the depletion in terms of volume. Irrigation will continue until the crops get matured. Volume of applied water can be calculated as follows:

$$d = V_i * D \tag{3.2}$$

$$V = 1000 * A * d \tag{3.3}$$

where:
d   = depth of application (m)
V_i = depleted water (Vol%)
D   = effective root depth (m)
V   = volume of water applied (lit)
A   = area of lysimeter (m^2)

Therefore, having the crop evaporation, crop coefficient value can be calculated as:

$$Kc = \frac{ETc}{ETo} \tag{3.4}$$

3.1.2 Modelling

Several crop models are available to simulate yield response to water. But they are used mostly by scientists, graduate students, and advanced users as these models require an extended number of variables and input parameters not easily available for the diverse range of crops and sites. Moreover, these models present substantial complexity and are rarely used by the majority of target users such as extension personnel, water user associations, consulting engineers, irrigation and farm mangers, planners and economists (Karunaratine et al., 2011). A relatively simple, water driven model, AquaCrop was used, as it emphasizes the fundamental process involved in crop productivity and in the response to water deficits.

AquaCrop is a water-driven model developed by FAO for use as a decision support tool in planning and scenario analysis in different seasons and locations (Steduto et al., 2009). Although the model is relatively simple and requires fewer data inputs, it emphasizes on the fundamental process involved in crop water productivity and in the response to water deficits, both from physiological and agronomic perspective. It is designed to balance simplicity, accuracy and robustness, and is particularly suited to address conditions where water is a key limiting factor in crop production. The key parameters are normalized water productivity, harvest index, canopy cover, yield and biomass (Steduto et al., 2009).

Models with empirical production functions are used as a practical option for the complexity of crop responses to water deficits. Among the empirical function approaches, using AquaCrop was based on the separation of: (i) the ET into soil evaporation (E) and crop transpiration (Tr) and; (ii) the final yield (Y) into biomass (B) and harvest index (HI). The separation of ET into E and Tr avoids the confounding effect of the non-productive consumptive use of water (E). This is important especially during incomplete ground cover. The separation of Y into B and HI allows the distinction of the basic functional relations between environment and B from those between environment and HI. The changes described led to the following equation at the core of the AquaCrop growth engine: $B = WP*Tr$; where Tr is the crop transpiration (in mm) and WP is the water productivity parameter (kg of biomass per m^2 and per mm of cumulated water transpired over the time period in which the biomass is produced). AquaCrop has a structure that overarches the soil-plant-atmosphere continuum (Steduto et al., 2009). It includes the soil, with its water balance; the plant with its development, growth and yield processes; and the atmosphere, with its thermal regime, rainfall and evaporative demand. Additionally, the management component has two main options, a field management option related to fertility level, mulching, etc. and a water management option related to rainfed / no irrigation / and irrigation. The irrigation option is particularly suited for simulating the crop response. A brief description of the input for the model is presented as follows.

Weather data collection

The atmospheric environment of the crop is described in the climate component of AquaCrop and deals with key input meteorological variables. Five weather input variables are required to run AquaCrop: daily maximum and minimum temperature, daily rainfall, wind speed, sunshine hours, daily evaporative demand of the atmosphere expressed as reference evaporation (ETo).

Crop data collection

The AquaCrop, the crop system has five major components: phonology, aerial canopy, rooting depth, biomass production and harvestable yield. The dates of the main phonological stages, canopy cover (CC), dry matter (DM), yield (Y) and harvest index (HI) were recorded. Plant height was measured in every ten days interval from the fixed sample of each experimental plot. Above ground biomass observations were made in every 10 days from an area of 1 m^2. The above ground biomass was dried with an oven drier for 48 h at 60 ^{0}C and then weighted. Canopy cover was measured by using LAI-2002 canopy analyzer from each treatment in every ten days interval. Grain yield was measured after maturity from pooled samples of an area of 1.5 m x 3.0 m in each plot. Crop management practices like sowing date, rate, method, cultivar, row spacing, weeding, fertilizer and pesticide application (type, amount, time and method) were also assessed.

Soil water data collection

To study the soil water balance, changes of incoming and outgoing water were measured for all plots both under rainfed and in irrigated fields in the experimental area of Melkassa Agricultural Research Centre.

A 1.0 x 1.0 m^2 pit was opened to take soil samples at different depths. Undisturbed soil samples were taken at a depth of 0-30 cm, 30-60 cm and 60-90 cm by using a core sampler with three replications. The initial soil moisture, bulk density, field capacity and permanent wilting point were determined. The field capacity was determined in the laboratory by using a pressure plate apparatus by applying a suction of -1/3 atmosphere to a saturated soil sample. Permanent wilting point was also determined in the laboratory as the moisture content corresponding to a pressure of -15 atmospheres from a pressure plate test.

Irrigation management

The water management considers options related to (i) rainfed agriculture (no irrigation), and supplementary irrigation (ii) irrigation no rain. Irrigation water was applied by using a hand held watering can having a fixed volume. Irrigation scheduling criteria were set based on the depth interval of irrigation and percentage of soil water content criteria. The irrigation option is particularly suited for simulating the crop response under supplemental or deficit irrigation.

The AquaCrop model has the potential to minimize the risks related to food insecurity in the country in general; because it can be used to explore and evaluate alternative management that improves water productivities and achieves more efficient water use. It might also be applied by extension specialists, relief organizations, and policy makers to predict yields.

4 Brief description of the Awash River Basin and the study area

4.1 The Awash River Basin

The Awash River Basin is the most important river basin in Ethiopia. The Awash River starts from the high plateau near Ginchi watershed in the central highlands of Ethiopia, flows along the Rift Valley into the Afar Triangle, and terminates in salty Lake Abbe on the border with Djibouti. The total length of the main course is 1,200 km. Based on physical and socio-economic factors the Awash River Basin is divided into upland (all lands above 1500 m+MSL), upper valley, middle valley (area between 1500 m and 1000 m+MSL), lower valley (1000 and 500 m+MSL) and eastern river basin (closed sub-basin area between 2500 and 1000 m+MSL). The upper, the middle and the lower valley are part of the Central Rift Valley systems. The Awash River Basin covers a total area of 110,000 km^2 of which 64,000 km^2 form the western river basin. They drain into the main river or its tributaries. The remaining 46,000 km^2 comprises the eastern river basins, drain into a desert area and do not contribute to the main river flow. The basin includes mainly the Afar, Oromiya and Amhara regions, including the area of the Addis Ababa city Administration and Dire Dawa that fall partly or wholly within the Awash River Basin, or have a significant proportion of their area falling inside the basin. Figure 4.1 presents the location of the Awash River Basin and the study area.

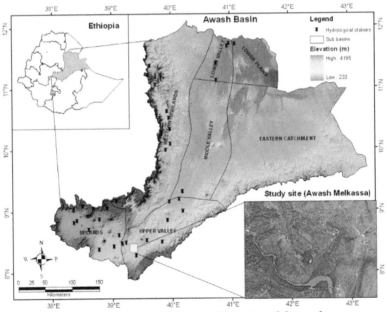

Figure 4.1. The Awash River Basin and location of the study area

A number of tributary rivers draining the highlands eastwards can increase the water level of the Awash River in a short period of time especially during August and September and cause flooding in the low-lying alluvial plains along the river course. Tributaries to Awash River such as Kessem, Kenena, Hawadi, Ataye Jara, Mille and Logiya rivers contribute most to the lowland flooding in Afar.

The irrigation potential for the Awash River is estimated to be 206,000 ha. But so far only 42.7% (88,000 ha) have been developed. Out of these 26.5% (23,306 ha) are under traditional and modern small-scale irrigation. The remaining 73.5% (64,694 ha) are developed under state farms and private investors. These include several agro-industries such as sugar factories and horticultural farms, ranches and cattle fattening, resort areas and other small industries.

The upper valley of the Awash River, where the study area is located, is the area between the Koka Dam and Awash station in which the river traverses some 300 km. The altitude ranges from 1000 - 2000 m+MSL; and annual rainfall varies from 600 - 800 mm. The dominant agriculture is grazing and irrigated cash crops.

4.2 Present conditions of Melkassa Agricultural Research Centre

Background information

Melkassa Agricultural Research Centre is one of the 12 federal research centres in Ethiopia. It focuses on dryland agriculture towards sustainability aspects of crop production and participation of stakeholders. Melkassa Agricultural Research Centre is situated in the semi-arid region of Ethiopia as 'hot to warm semi-arid lakes and Rift Valley'. Mamo (2006) divided the Rift Valley in four agro climatic zones depending on rainfall amount and risk assessment in crop production. Melkassa is categorized under 3 zones characterized by an annual rainfall ranging between 600 - 800 mm. This zone is characterized as being relatively wet with 75% of the years getting optimum rainfall.

Climate

Melkassa is classified in the semi-arid agro ecological zone by Ethiopia Ministry of Agriculture (MOA) (2000) and Bennie and Hensley (2001). The monthly rainfall distribution is a bimodal type of rainfall distribution. The main rainy season is during the months of June to September, during which 68% of the annual rainfall occurs (Table 4.1). The highest evaporative demand occurs during the months of March, April and May. During these months the mean maximum temperature (Tmax) is around 30 ^{0}C while the mean relative humidity (RHm) drops to 51%. During the main crop growing season of June to September conditions are more favourable with Tmax and RHm approximately 27 ^{0}C and 64% respectively. The agro ecological zones classification of Ethiopia (Ministry of Agriculture (MOA), 2000; Engida, 2000) mentions that Melkassa is classified under areas with two growing seasons of 50 and 100 days length for the 1st and 2nd seasons, respectively, and has an annual rainfall and potential evapotranspiration of about 772 mm and 1994 mm, respectively. The aridity index (AI) of 0.39 identifies this as a semi-arid area. Table 4.1 presents the mean monthly climatic data and the values of AI.

Soil

The Soil is classified as a Hypo Calcic Regosol (WRB Classification). The ecotope name is therefore Melkassa Calcic Fluvic Regosol. This soil covers about 10% of Ethiopia and about 16% of the Rift Valley (Itanna, 2005). The important characteristics of the Melkassa Soil are a favourable clay loam texture of the fine earth throughout the profile, with high silt content with effective depth of 100 - 150 cm. The top soil is strongly crusting. The water holding capacity of such soils within the root zone is considered to be high.

Table 4.1.Mean monthly climatic data of Melkassa Agricultural Research Centre (1977-2012)

Month	Rainfall (mm)	Min T (°C)	Max T (°C)	RH (%)	Sunshine (h)	Windspeed (km h^{-1})	ETo (mm)	AI
January	14	12	28	52	8.9	11	167	0.08
February	26	13	29	50	8.7	12	167	0.16
March	51	15	30	52	8.3	11	189	0.27
April	52	15	30	51	8.3	10	180	0.29
May	52	16	31	51	8.9	10	186	0.08
June	68	16	30	53	8.4	12	177	0.38
July	186	16	27	67	7.0	12	149	1.25
August	181	15	26	69	7.2	10	140	1.29
September	82	14	27	65	7.3	6	135	0.61
October	42	12	29	50	8.6	8	164	0.26
November	8	11	28	46	9.7	11	171	0.05
December	11	11	28	49	9.5	11	171	0.06
Total	772						1994	
Mean		14	29	55	8.4	10		0.39

ETo = Reference Evapotranspiration, AI = Aridity Index

4.3 Aridity index (AI)

Definition of aridity

Aridity is a term that most people conceptually understand, and it evokes images of dry, desert lands with sparse natural surface water bodies and rainfall, and commonly only scant in vegetation, which is adapted to a paucity of water. Aridity largely occurs on arid and semi-arid lands with warm climate in which water scarcity is more severe because of greater population and associated water use. However, aridity also occurs in regions with cold climates in which precipitation falls only in snow. A fundamental distinction exists between aridity, which is a long-term climatic phenomenon and droughts, which are temporary phenomena (water deficit) (Stadler, 2005).

Aridity indices

The simplest AI is based solely on precipitation. The commonly used rainfall based definition is that an arid region receives less than 250 mm of precipitation per year (Intergovernmental Panel on Climate Change (IPCC), 2007). Semi-arid regions are commonly defined by annual rainfall between 250 and 500 mm). Figure 4.2 presents the regionally classified AI map of Ethiopia.

The United Nation Environmental Programme (UNEP) (1992) AI is based on the ratio of precipitation (P) and potential evapotranspiration as follows (Equation 4.1):

$$AI = \frac{P}{ET_p} \qquad (4.1)$$

where:
AI = aridity index
P = average annual precipitation (mm)
ET_p = average annual potential evapotranspiration (mm)

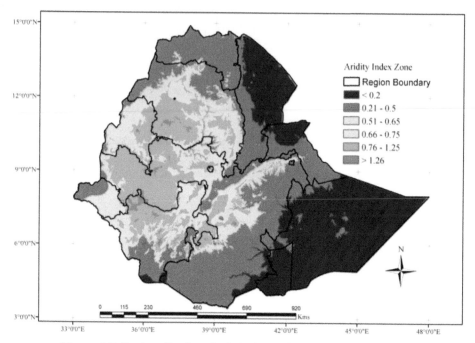

Figure 4.2. Regionally classified Aridity Index (AI) map of Ethiopia

Warm arid regions have low P and high ET_p rates and thus very low AI values. UNEP (1992) proposed a classification of climate zones based on AI index, in which arid regions are defined by an index of less than 0.2 (Table 4.2). Alternative versions of the classification use an AI value for 0.05 for the boundary between hyper-arid and arid regions.

Table 4.2.Aridity Classification (UNEP, 1992)

Classification	Aridity Index
Hyper-arid	AI < 0.050
Arid	0.05 < AI < 0.20
Semi-arid	0.20 < AI < 0.50
Dry subhumid	0.05 < AI < 0.65

5 Crop water productivity of irrigated Teff in a water stressed region

5.1 Introduction

The dryland areas of Ethiopia account for more than 70% of the total land mass and 40% of the arable land. However, these areas contribute less than 30% of the country's total agricultural production. The country receives rainfall, but seasonal and concentrated in three months of the year from June to August. The traditional rainfed agriculture concentrated in the highlands appears to shoulder the responsibility of feeding the human population exceeding 77.2 million (Central Statistical Agency (CSA), 2007). Thus food insecurity has remained to be the major problem that is a great concern to the country. Therefore it is imperative to bring large areas of the arid, semi-arid and sub-humid regions with uneven rainfall distribution under irrigation.

Irrigation has a multi-faceted role in contributing towards food security, self-sufficiency, food production and exports (Hussain and Hanjra, 2004). Scarce water resources and growing competition for water will reduce its availability for irrigation. At the same time, the need to meet the growing demand for food will require increased crop production from less water (Ali and Talukder, 2008). Achieving greater efficiency of water use will be a primary challenge in the near future and will include the employment of techniques and practices that deliver a more accurate supply of water to crops.

Increasing crop water productivity (CWP) as argued by FAO (2010) and Geerts and Raes (2009) can be an important pathway for poverty reduction. This would enable growing more food and hence feeding the ever increasing population of Ethiopia or gaining more benefits with less water thus enhancing the household income. CWP can be improved by proper irrigation scheduling, which is essentially governed by crop evapotranspiration (ETc). Therefore it is profoundly important to determine and know the crop water requirement, the crop coefficient and the yield response factor of Teff crop.

Teff is the major indigenous cereal crop of Ethiopia. Teff flour is primarily used to make a fermented, sour dough type, flat bread called Injera. Teff is also eaten as porridge or used as an ingredient of home-brewed alcoholic drinks. It also has a high iron content and high potential as an export crop to USA and European countries as it contains no gluten and is considered a healthy food grain (Roseberg et al., 2006). Approximately 1 million Americans suffer from celiac disease (gluten sensitivity) and Teff may provide a niche for meeting the dietary requirement (Spaenij-Dekking et al., 2005). Serious attempts are underway to expand its cultivation in Europe, notably in the Netherlands and the United States of America (Evert et al., 2009).

Teff is a highly demanded cereal and has higher market prices than the other cereals for both its grain and straw in Ethiopia. Farmers earn more for growing Teff than growing other cereals. More than half of the area under cereals in Ethiopia is for Teff production. Teff grain is not attacked by weevils, which means that it has a reduced postharvest loss in storage and requires no pest-controlling storage chemicals. Habtegebrial and Singh (2006) investigated the impact of tillage and Nitrogen fertilizer on yield of Teff. Many investigations have been carried out and indicate that Teff is adapted to environments ranging from drought-stressed to waterlogged soil conditions (Roseberg et al., 2006). Despite of this fact, with respect to Teff as a food crop in Ethiopia, where it originated and was diversified (Assefa et al., 2003), there has been only limited research on its agronomic and physiological responses to water and other physical stress. Although (Mengistu, 2009) studied the physiological responses of Teff to water stress in greenhouse conditions its degree of tolerance for specific levels of water application at field scale has not yet been investigated. Hence, it is important to examine the response of Teff to full and limited irrigation conditions at field scale.

5.2 Experimental site

The field experiments were conducted at Melkassa Agricultural Research Centre. The research centre is located in the Central Rift Valley at 8°24′N latitude, 39°21′E longitude, and altitude of (1,300 – 1,800 m+MSL). The area is among the semi-arid regions and characterized by erratic rainfall, frequent drought and a harsh cropping environment. The mean minimum and maximum monthly temperatures of the area are 22 °C and 34 °C respectively. Teff is one of the common crops grown in the area that is considered as a reliable and low-risk crop. It is grown during the summer rainy season from June to August. The texture of the soil is Clay Loam.

5.3 Climatic data collection and analysis

Climate data including daily rainfall, maximum and minimum temperature, relative humidity, sunshine hours and wind speed were obtained from the meteorological station near the experimental field. The ETo calculator was used to determine the daily reference evapotranspiration (ETo) for the growing season of 2010/2011 and 2011/2012. ETo calculator is software developed by the Land and Water Division of FAO. Its main function is to calculate Reference evapotranspiration (ETo) based on computation guidelines detailed in (Raes, 2009).

5.4 Field experiments

The field experiments were carried out in the dry season of 2010/2011 and 2011/2012. A selected combination of depth of irrigation water application (amount) and growth stage (time) of Teff (*Eragrostic Tef*) was used as experimental design in order to determine the optimum water application depth at specific growth stages that results in optimum crop water productivity (CWP). This research investigated the sensitivity of each growing stage to drought stress in detail. Four different levels of irrigation water supply were scheduled, full crop water requirement 0% deficit (ETc), 25% deficit (applying 75% of crop water requirement), 50% deficit (applying 50% of crop water requirement) and 75% deficit (applying 25% crop water requirement).

In Figure 5.1, T_1 to T_{16} refers to different treatments (crop stands) under various combinations of four growth stages (I to IV) and irrigation applications starting from no deficit (0%D) to the maximum of 75% deficit (75%D).

The phenological cycle was divided into phases which are considered to be most relevant from the viewpoint of their response to irrigation, i.e. initial stage (P_1), development stage (P_2), mid season stage (P_3) and late season stage (P_4). A four by four factorial combination of sixteen treatments with three replications was set in the experimental field to make a total of forty-eight trials (Figure 5.1). Each set of these 48 trials was tested at seeding rates of 10 and 25 kg/ha. Thus the total field experimental plots established in Melkassa Agricultural Research Centre were 96.

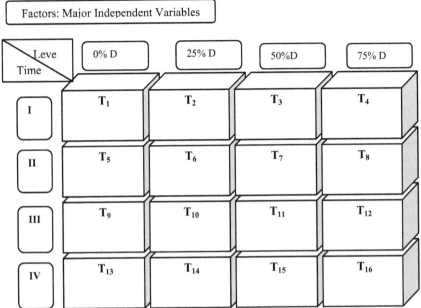

Figure 5.1. Field experimental set-up for assessing crop sensitivity to different water application scenarios

5.5 Soil and cultivar

Soil samples were collected depending on the root depth of the experimental crop Teff at the time of sowing. The physical soil characteristics are given in the Table 5.1.

Table 5.1. Soil physical characteristics of the experimental field

Soil property	Soil depth (cm)		
	0-30	30-60	60-90
Particle size distribution (%)			
Sand (> 50 µm)	37.8	38.3	36.8
Silt (2-50 µm)	35.0	36.6	39.7
Clay (<2 µm)	27.3	25.1	23.6
Textural class	Clay loam	Loam	Loam
Bulk density (g/cm^3)	1.4	1.5	1.8
Field capacity (FC) (Vol%)	29.3	31.6	34.4
Permanent wilting point (PWP) (Vol%)	16.7	18.4	21.3
Total available water (TAW) (mm/m)	125.6	132.2	141.5

Teff (*EragrosticTef*) cultivar, *Kuncho*, was selected and its CWP was assessed under the 16 different treatments as outlined in Figure 5.1. The assessments were done based on actual grain and biomass yields obtained during two irrigation seasons: 1) November 2010 to March 2011; and 2) December 2011 to April, 2012. *Kuncho* was selected because as compared to the other local varieties released by Melkassa Agricultural Research Centre, it is the most favourite among the locals and has high market value within the country and the region.

Each treatment had three replicates thus a total of 48 experimental fields were analysed. Each experimental field has an area of 8 m² (2.0 m x 4.0 m). The individual fields were separated from each other by means of soil bunds

Irrigation was applied manually using a watering can with a known capacity. In line with the rate of applications of fertilizers in the research area, 130 kg/ha of di-amonium nitrogen phosphate (DAP) was spread in each field just before planting. Likewise 35 kg/ha of urea was applied at tillering stage.

The dates of the main phenological stages, like sowing date, date of 90% emergence, 50% flowering, duration of flowering, senescence and maturity were recorded. Plant height was measured in every ten days interval from the fixed sample of each experimental plot. Above ground biomass observations were made in every 10 days from undisturbed sample areas of 1 m². The above ground biomass was dried with an oven drier for 48 hours at 60 °C and then weighted. Grain yields were measured after maturity from pooled samples of an area of 2.0 x 3.0 m² in each plot. The crop was harvested manually. The grain and total biomass fresh weight were weighted at maturity and then were dried and weighted on a sensitive balance.

5.6 Soil moisture monitoring

Soil moisture was regularly monitored using a Neutron Probe. Irrigation was applied in accordance with the different water deficit conditions. In the case of a full irrigation application (no deficit), the soil moisture was constantly kept at field capacity. The Neutron Probe was calibrated by using the Gravimetric Method whereby soil samples were collected from soil depths of 15, 30 and 45 cm before sowing and in every 5 - 7 days interval up to maturity. The measured soil moisture in weight basis was multiplied by bulk density to convert to volume basis.

5.7 Seasonal evapotranspiration calculation using soil-water balance approach

Evapotranspiration (ET) was calculated on the basis of the general water balance equation (Equation 5.1):

$$P + I + U = D + R \pm \Delta W + ETa \tag{5.1}$$

where:
P   = effective rainfall (mm)
I   = irrigation water applied (mm)
U   = upward flux (mm)
D   = deep percolation below the crop root zone (mm)
ΔW = change in soil moisture storage in the soil profile (mm)
ETa   = crop evapotranspiration (mm)

In these experiments, as the field was sufficiently levelled, water was applied only to refill to field capacity and while the groundwater table was far below the root zone D and U were neglected. Surface runoff was assumed to be zero as the irrigation water was protected by constructed soil bunds around each plot. The reduced soil water balance equation becomes:

$$ET = I + P \pm \Delta W \tag{5.2}$$

5.8 Crop water productivity (CWP)

Crop water productivity (CWP) is defined by different definitions of various researchers (Bessembinder et al., 2005). In this chapter, CWP is defined as the amount or the value of product over volume or value of water depleted or diverted.

CWP is computed as the ratio of grain yield to actual crop water use:

$$CWP = Y / ETa \qquad\qquad (5.3)$$

where:
CWP = expressed in kg/m^3 on a unit water volume basis
Y   = grain yield (kg/ha)
ETa   = actual crop evapotranspiration (m^3/ha)

5.9 Statistical analysis

The statistical difference in Teff yield and biomass for different treatments was analysed using MSTAT-C statistical package for Analysis of Variance (ANOVA). MSTAT-C is a computer based statistical software package developed by the Crop and Soil Science Department. The Least Significant Difference (LSD) test was performed at Alpha level of 1% (highly significant) and 5% (significance) level.

5.10 Results and discussion

5.10.1 Effect of moisture level on grain yield and above ground biomass yield of Teff

The seasonal total rainfall from June to August in 2010 and 2011 was 555 mm and 437 mm respectively. The long-term average growing season of rainfall was 768 mm. But for the dry season irrigation experiment no effective rainfall was recorded. The irrigation amounts for different treatments are given in Table 5.2.

The effect of moisture level on grain yield for different seeding rates is shown in Figure 5.2. Using the Least Significant Difference (LSD) test, first confirming that the F-test was significant at 5% and 1% significance level, it was found that the effect of different levels of moisture on Teff grain yield shows a highly significant ($p < 0.01$) and significant ($p < 0.05$) difference. Moreover, the above ground biomass yield has resulted in highly significant difference with the different levels of moisture. The water deficit at the initial stage and late season stage for both 75% deficit and 50% deficit, gave non-significantly ($p > 0.05$) different yields from the optimum application of treatment ET1 (100%). However, for all levels of water deficit at the mid season stage and 50% and 75% deficits throughout the growth stages the yields were significantly different ($p < 0.01$) as compared to 0% deficit.

In Table 5.2 T_1 to T_{16} refer to different treatments (crop stands) under various combinations of four growth stages (I to IV) and irrigation applications starting from no deficit (0%D) to the maximum of 75% deficit (75%D).

In the experimental season of 2010/2011, irrigation was applied every three days interval, more frequent but less water. But in the experimental season of 2011/2012, a comparatively higher amount of irrigation water was applied in six days interval.

Table 5.2. Irrigation application under different treatments (water deficit conditions) for both experimental seasons

Treatments	2010/2011		2011/2012	
	No of irrigation (nos)	Irrigation amount (mm)	No of irrigation (nos)	Irrigation amount (mm)
T_1 (0%D)	15	281	10	380
T_2 (25%D)	15	211	10	294
T_3 (50%D)	15	141	10	225
T_4 (75%D)	15	70	10	156
T_5 (75%D)I	15	271	10	325
T_6(75%D)II	15	238	10	354
T_7(75%D)III	15	163	10	350
T_8(75%D)IV	15	241	10	367
T_9(50%D)I	15	274	10	307
T_{10}(50%D)II	15	252	10	332
T_{11}(50%D)III	15	202	10	359
T_{12}(50%D)IV	15	254	10	352
T_{13}(25%D)I	15	278	10	294
T_{14}(25%D)II	15	267	10	315
T_{15}(25%D)III	15	242	10	318
T_{16}(25%D)IV	15	267	10	349

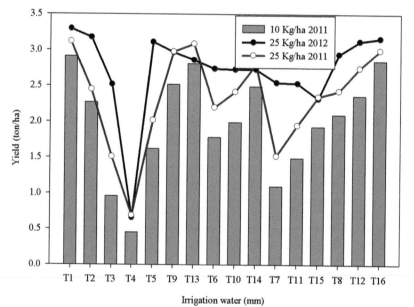

Figure 5.2. Yield of Teff under different irrigation treatments for the experimental seasons of 2010/2011 and 2011/2012

5.10.2 Grain yield, crop water productivity (CWP) and relative yield decrease

The main grain yield, crop water productivity and relative yield reduction for the different treatments for the 2010/2011 experimental years are presented in Table 5.3. The response of Teff to various levels of irrigation water is different. During 2010/2011, when 100% of ETc was applied (0% deficit - treatment T_1 in Table 5.3), grain yield of

Teff for the seeding rate of 25 kg/ha and 10 kg/ha was found to be 3.12 and 2.91 ton/ha respectively. In case of treatment T_2 (75% of ETc irrigation application i.e. 25% deficit) the yields values were reduced to 2.45 and 2.27 ton/ha respectively. Much more significantly lower yields of 0.69 ton/ha and 0.45 ton/ha were obtained for treatments with 75% irrigation deficit (T_4) throughout the whole growth stage for both 25 kg/ha and 10 kg/ha seeding rates respectively.

It can be inferred from Table 5.3 that 75, 50 and 25% irrigation water reduction throughout the whole growth stage decreased the Teff yield by 77.9, 51.6 and 21.5% respectively. These results are in agreement with Kloss (2012) who showed that dealing with improvement of WP is closely related to the irrigation practice of regulated deficit irrigation and has a direct effect on yield as the amount of water applied decreases intentionally the crop yield drops.

In Table 5.3, T_1 to T_{16} refer to different treatments (crop stands) under various combinations of four growth stages (I to IV), irrigation applications starting from no deficit (0%D) to the maximum of 75% deficit (75%D) and SDR refers to seeding rate. Treatment means with ** and * refer to highly significant (P < 0.01), significant (P < 0.05) and NS non significant variation.

Decreasing the recommended seeding rate of 25 kg/ha to 10 kg/ha increases the capacity of the Teff stalk to increase its strength and be able to have a greater resistance of the stem against lodging because of heavy panicle at maturity. In addition smaller seeding rate increases the tillering potential. For seeding rate of 10 kg/ha, a maximum yield of 2.91 and 3.05 ton/ha respectively was obtained from the treatment which received the optimum crop water requirement (T_1) during the 2010/2011 and 2011/2012 experimental seasons. Generally regardless of the seasons, the patterns of response to irrigation treatment were similar and showed significant and positive response. As the price of Teff seed increased 10 fold, decreasing the seeding rate can save seed to cultivate additional ha of land.

Higher crop water productivity values of 1.16 and 1.08 kg/m^3 were obtained for seeding rate of 25 kg/ha and 10 kg/ha, respectively from the treatment which received 75% ETc throughout the growth stage in the experimental season of 2010/2011. Whereas for the experimental season of 2011/2012 higher crop water productivity values of 1.12 and 1.31 kg/m^3 were obtained with seeding rates of 25 kg/ha and 10 kg/ha respectively (Table 5.4). Katerji et al. (2010) observed that CWP varied with years and locations and ranged from 1.34 to 1.81 kg/m^3, which were similar to the results reported by Zhang and Yang (2004) with CWP ranging between 1.01 and 1.72 kg/m^3.

In Table 5.4, T_1 to T_{16} refer to treatments (crop stands) under various combinations of four growth stages (I to IV) and irrigation applications starting from no deficit (0%D) to the maximum of 75% deficit (75%D). Again treatment means with ** and * refer to highly significant (P < 0.01), significant (P < 0.05) and NS non significant variation.

Higher crop water productivity values of 1.16 and 1.08 kg/m^3 were obtained for seeding rate of 25 kg/ha and 10 kg/ha, respectively from treatment which received 75% ETc throughout the growth stage in the experimental season of 2010/2011. Whereas for the experimental season of 2011/2012 higher crop water productivity value of 1.12 and 1.31 kg/m^3 were obtained with seeding rate of 25 kg/ha and 10 kg/ha respectively (Table 5.4). Katerji et al. (2010) observed that CWP varied with years and locations and ranged from 1.34 to 1.81 kg/m^3, which were similar to the results reported by Zhang and Yang(2004) with CWP ranging between 1.01 and 1.72 kg/m^3.

Table 5.3. Effect of moisture levels and seeding rate on grain yield and crop water productivity of Teff at growing season 2010/2011

*Treatment Name	Irrigation (mm)	Grain yield (kg/ha) at SDR of		Crop water productivity (CWP) at SDR of		Relative grain yield reduction at SDR of		Relative evapotranspiration deficit at SDR of	
		25 kg/ha	10 kg/ha	25 kg/ha	10 kg/ha	25 kg/ha	10 kg/ha	25 kg/ha	10 kg/ha
T_1 (0%D)	281	3,120	2,910	1.11	1.04	0	0	0	0
T_2 (25%D)	211	2,450*	2,270	1.16	1.08	0.21	0.22	0.25	0.25
T_3 (50%D)	140	1,510**	960	1.08	0.68	0.52	0.67	0.50	0.50
T_4 (75%D)	70	690**	450	0.98	0.64	0.78	0.85	0.75	0.75
T_5 (75%D)I	271	2,020*	1,620	0.75	0.60	0.35	0.44	0.04	0.04
T_6 (75%D)II	238	2,200*	1,780	0.92	0.75	0.29	0.39	0.15	0.15
T_7 (75%D)III	163	1,520**	1,100	0.93	0.67	0.51	0.62	0.42	0.42
T_8 (75%D)IV	241	2,430*	2,100	1.01	0.87	0.22	0.28	0.14	0.14
T_9 (50%D)I	274	2,980NS	2,520	1.09	0.92	0.04	0.13	0.02	0.02
T_{10} (50%D)II	252	2,420*	1,990	0.96	0.79	0.22	0.32	0.10	0.10
T_{11}(50%D)III	203	1,950**	1,490	0.96	0.74	0.38	0.49	0.28	0.28
T_{12} (50%D)IV	254	2,750NS	2,360	1.08	0.93	0.12	0.19	0.10	0.10
T_{13} (25%D)I	278	3,090NS	2,810	1.11	1.01	0.01	0.03	0.01	0.01
T_{14} (25%D)II	267	2,800NS	2,490	1.05	0.93	0.10	0.14	0.05	0.05
T_{15} (25%D)III	242	2,350*	1,930	0.97	0.80	0.25	0.34	0.14	0.14
T_{16} (25%D)IV	267	3000NS	2,850	1.12	1.07	0.04	0.02	0.05	0.05

Table 5.4. Grain yield, biomass yield, crop water productivity (CWP) and relative yield decrease (RYD) by treatment for 25 kg/ha for the experimental season of 2011/2012

Treatments	Yield (ton/ha)	Biomass (ton/ha)	Irrigation applied (mm)	CWP (kg/m³)	RYD relative yield reduction	Relative ET deficit
T₁ (0%D)	3.30	14.0	380	0.87	0.00	0.0
T₂ (25%D)	3.18NS	11.2*	294	1.08	0.04	0.2
T₃ (50%D)	2.52*	6.7**	225	1.12	0.24	0.4
T₄ (75%D)	0.66**	4.5**	156	0.42	0.80	0.6
T₅ (75%D)I	2.86NS	6.2**	294	0.98	0.13	0.2
T₆ (75%D)II	2.73NS	8.2NS	315	0.87	0.17	0.2
T₇ (75%D)III	2.32*	8.0*	318	0.73	0.30	0.2
T₈ (75%D)IV	3.17NS	11.0NS	349	0.91	0.04	0.1
T₉ (50%D)I	2.97NS	6.7**	307	0.97	0.10	0.2
T₁₀ (50%D)II	2.74*	9.0**	332	0.82	0.17	0.1
T₁₁ (50%D)III	2.54*	12.0NS	359	0.71	0.23	0.1
T₁₂ (50%D)IV	3.13NS	11.3NS	351	0.89	0.05	0.1
T₁₃ (25%D)I	3.11NS	8.8**	325	0.96	0.06	0.1
T₁₄ (25%D)II	2.77*	10.0NS	354	0.78	0.16	0.1
T₁₅ (25%D)III	2.55*	11.5NS	350	0.73	0.23	0.1
T₁₆ (25%D)IV	2.94NS	11.7NS	367	0.80	0.11	0.0

During 2011/2012, grain yields and biomass followed a similar trend to those of 2010/2011. Treatments T_7 (75%D) III, T_{11} (50%D) III and T_{15} (25%D) III which were conducted under adequate watering conditions throughout the first two periods of the growing season, and followed by a period of stress at the mid season stage with 75%D, 50%D and 25%D resulted in the second, the third, and the fourth lowest yield respectively by 30, 23 and 23% for the seeding rate of 25 kg/ha. This yield reduction is significant compared with stressing the crop during late season stage having a reduction of 11%.

This tendency might be attributed to the fact that adequate watering conditions early in the season led to the development of an abundant leaf cover and a shallow root depth. When a severe stress follows, the crop rapidly depletes the soil water stored in the root zone and wilts before completion of additional root development at greater soil depths. This result is in agreement with the irrigation experiment of grains that are Sorghum and Maize, showed that the extent of yield reduction following evapotranspiration deficit depends on the growth period (Stewart et al., 1975). From deficit irrigation experiments on vegetables and cereals, it was found that lowest yield is obtained during the full stress (75% deficit) throughout the growing season. However, stressing the crops during initial and late season stage does not affect the crop yield significantly.

Figure 5.3 shows the graphical relationship between the relative yield and biomass reduction with moisture stress for the experimental season of 2010/2011. The crop response factor (K_y), the slope of total grain yield and above ground biomass reduction versus moisture stress graph was found to be 1.09 and 1.19 respectively for 25 kg/ha and 10 kg/ha seeding rate and 0.88 and 0.96 respectively for 25 kg/ha and 10 kg/ha seeding rate. The relative biomass reduction increases as the evapotranspiration deficit increases. Here K_y for biomass yield was found to be 0.88 which is less than unity in contrast with the yield response factor of grain yield.

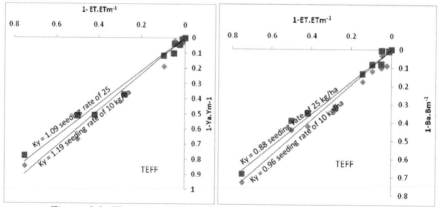

Figure 5.3. Water production functions of Teff yield and biomass

$K_y < 1$ indicates that biomass production is less sensitive to water stress. According to FAO (2002) only those crops and growth stages with a lower yield response factor ($K_y < 1$) can generate significant savings in irrigation water through deficit irrigation. A response factor greater than unity indicates that the expected relative yield decrease for a given evapotranspiration deficit is proportionately greater than the relative decrease in evapotranspiration. This means that advantages coming from deficit irrigation are unlikely (Kirda, 2002). This implies that Teff is sensitive to water stress conditions. As shown in Table 5.3 and 5.4, the yield has decreased from

3.12 to 0.69 ton/ha and 3.30 to 0.66 ton/ha respectively in the experimental years of 2010/2011 and 2011/2012. On the other hand, as presented in Table 5.3, for the higher seeding rate of 25 kg/ha, respective of higher water productivity of 1.16 and 1.08 kg/m^3 was obtained from treatments which received 25% and 50% deficit irrigation throughout the whole growth season. Whereas for the lower seeding rate of 10 kg/ha, higher water productivity of 1.08 and 1.07 kg/ha was obtained from the treatment which received 25% deficit irrigation throughout the growth stage and 25% deficit during the late season stage respectively. By growing more with less water, more water will be available to irrigate arable land in a water scarce semi-arid region of Ethiopia.

5.11 Concluding remarks

The following major conclusions could be drawn:
- a maximum grain yield of 3.3 ton/ha was obtained under irrigation when Teff was no subject to any water stress. This is three fold the yield farmers currently harvest from rainfed agriculture;
- the crop was very sensitive to water stress during the mid season stage. Even when it is only subject to 25% deficit, its yield decreased by about 1 ton/ha. Stressing the crop either by one-half or three-quarters at the mid season stage, results in lower yields next to stressing the crop throughout the growth season;
- the yield and water productivity differences are insignificant between full irrigation and 25% deficit irrigation distributed throughout the growth period at seeding rates of 25 kg/ha and 10 kg/ha. Thus, when water is scarce and irrigable land is relatively abundant as is the case in Ethiopia, adopting the 25% water deficit irrigation with 10 kg/ha seeding rate is recommended;
- a maximum water deficit of 50% during the late season stage has an insignificant impact on Teff yield and water productivity.

6 Teff (*Eragrostic Tef*) crop coefficient for effective irrigation planning and management in the semi-arid Central Rift Valley of Ethiopia

6.1 Introduction

Teff is a staple food grain for more than 85% of the 90 million population of Ethiopia and covers 75% of cultivated land in the country. Farmers and commercial growers produce Teff for both local and export markets. For centuries and until the last 5 years, Teff has only been grown as a rainfed crop and produced only once in a year with a maximum yield of 1 ton/ha. As the price of Teff increased fivefold over the last decade from 2,000 Birr (about US$ 130) per ton to 10,000 Birr (US$ 650) per ton (Famine Early Warning System Network Ethiopia (FEWSNET), 2012), increasing the production of the crop including using irrigation has become important. However, in arid and semi-arid regions such as the Central Awash Rift Valley, water is scarce and maximizing crop and water productivity requires proper irrigation planning and management. Accurate determination of Crop coefficient (Kc), the ratio between potential crop evapotranspiration (ETc) and reference evapotranspiration (ETo) is a key factor for effective irrigation planning and management.

Kc is the ratio of crop evapotranspiration (ETc) to reference crop evapotranspiration (ETo) that can be established based on crop evapotranspiration and climate (Allen et al., 1998). It is a function of climate, crop type, crop growth stages, soil moisture and irrigation method (Kang et al., 2003). Kc can be expressed with respect to the days after sowing (DAS), which helps to relate Kc to the different crop development stages. For most crops, values of Kc increase from a minimum value at planting and a maximum value at full canopy cover and finally decline as the plant reaches late season stage (Allen et al., 1998; Kashyap and Panda, 2001; Sepaskhah and Andam, 2001b)

Crop evapotranspiration is a major component of the agricultural water budget and is a key factor to determine a proper irrigation schedule and improve water productivity in irrigated agriculture (Irmak et al., 2008; Payero et al., 2008), especially in arid and semi-arid areas where supplementary irrigation is necessary to minimize risk of crop failure. Crop evapotranspiration can be determined experimentally using lysimeters by monitoring the crop root zone water flux over the growing period (Allen et al., 1998). Jensen (1974) introduced the concept of crop coefficient (Kc) and it was further developed by other researchers (Doorenbos and Pruitt, 1975, 1977; Burman et al., 1980a; Allen et al., 1998).

Several studies have been conducted over the years to measure Kc for different crops. Since most of the studies have been specific to one or two crops, FAO prepared a comprehensive list of Kc values for various crops under different climatic conditions by compiling results from different studies (Doorenbos, 1979; Allen et al., 1998). Researchers have emphasized the need for regional calibration of Kc under given climatic conditions for a realistic estimation of crop water requirement (Kang et al., 2003). Field experiments based accurately measured Kc values for irrigated Teff grown in the semi-arid region of Central Rift Valley of Ethiopia are not available.

As part of the research, field experiment were conducted in the period 2010 to 2012 to determine the Kc, crop evapotranspiration and yield values of local Teff variety. This chapter presents and discusses the research approach, the main results of the field experiments and accordingly draws some conclusions and recommendations.

6.2 Methods used

6.2.1 The experimental crop - Teff

Teff cultivar *Kuncho* (Dz-Cr-387-RIR-355) was sown during four seasons on 7 December, 2010 (Lysimeter 1); 7 March, 2011 (Lysimeter 2); 7 December 2011 (Lysimeter 3) and 11 March, 2012 (Lysimeter 4) at a seeding rate of 25 kg/ha both inside and surrounding (buffer area) of the Lysimeters. Fertilizer was applied according to the recommended rate for the area (130 kg/ha DAP at planting and 35 kg/ha urea at tillering).

6.2.2 Crop coefficient (Kc)

As stated before, in this study the crop coefficient was calculated from estimated crop evapotranspiration by field water balance method and reference evapotranspiration by using software called ETo calculator.

Factors determining Kc consist of climate conditions and crop growth stages. The crop growth period for Teff were classified into four distinct growth stages as initial, crop development, mid season and late season stages. Even though the length of growing stages of Teff with respect to climatic region is not available in FAO (Allen et al., 1998), the length of the growing stage available in FAO for Millet and Barley (having similar characteristics with Teff) was used (Germa and Hess 1994). Two years experimental field research were undertaken to determine Kc. The following equation was used to calculate the Kc:

$$Kc = \frac{ETc}{ETo} \qquad (6.1)$$

where:
Kc   = crop coefficient
ETc   = crop evapotranspiration (mm)
ETo   = reference evapotranspiration (mm)

6.2.3 Relationship of crop coefficient (Kc) and days after sowing (DAS)

The variations in Kc as a function of days after sowing (DAS) were calculated. A multiple regression analysis equation was developed to determine the relationship of Kc and days after sowing (DAS).

6.2.4 Setup of lysimeters at Melkassa Agricultural Research Centre

In Ethiopia Melkassa Agricultural Research Centre is one of the research centres where a well-functioning lysimeters are available (Figure 6.1). Totally eight non-weighing lysimeters with one big drainage pit were available to control and measure the drainage from each lysimeter. The lysimeters were located near (100 m) from the agro-meteorological station of the research centre. The lysimeters were of the non-weighing type, two of them are square in cross-section with 2 m length, 2 m width and 2 m depth and the other two are rectangular with 2 m length, 1 m width and 2 m depth. They were constructed of reinforced concrete and the inside was lined with plastic sheet to avoid leakage or lateral in and out flow of water. The lysimeters were filled with soil from nearby locations maintaining the natural vertical sequence of the soil. They have an access chamber (for aeration) and underground steel pipes were used to drain water into graduated buckets. The height of each lysimeter rim was maintained near the ground level to minimize the boundary layer effect around the lysimeter. Rim of each lysimeter protrude 10 cm above the soil surface so that no surface runoff water could enter into the lysimeters. In the centre of each lysimeter, one access tube for soil moisture meter was installed up to the depth of 90 cm to enable accurate determinations of the soil water content.

Figure 6.1. The lysimeters

6.2.5 Crop evapotranspiration (ETc)

The stored soil moisture, which is one of the important inputs for the calculation of crop evapotranspiration (ETc), was monitored by a Neutron Probe on every alternate day. For greater accuracy, the Neutron Probe was calibrated for the soil type of the experimental plots using the Gravimetric Method by establishing wet and dry points, to obtain a wide range of moisture and to make it possible for the probe to read the ranges. As the instrument was not consistently accurate in measuring the upper 0-15 cm depth, this was monitored using the Gravimetric Method. Probe readings were taken for the two depth intervals, 15-30 cm and 30-45 cm. The maximum depth of soil moisture measurement (45 cm) corresponded to the depth of root zone of Teff. The amount of irrigation water applied depended on weather, crop growth stage and rooting depth. Irrigation was applied to the crop when about 55% of the available soil moisture was depleted from the effective root zone as recommended by Allen et al. (1998). The depletion was converted into volume basis which was manually applied using a watering can with a known volume.

The amount of water entering and leaving each lysimeter was recorded and used to determine the ETc during the growing season. The average daily ETc in a certain period of time was determined using the following water balance equation:

$$ETc + Ro + D = I + Pe + Cf \pm \Delta S_{105} \tag{6.2}$$

where:
ETc = crop evapotranspiration (mm)
Ro   = surface runoff (mm)
D   = drainage water collected and measured (mm)
I   = irrigation (mm)
Pe   = effective rainfall (mm)
Cf   = upward flux from the groundwater table (mm)
ΔS = change in soil water content in the root zone for a given time interval in the layer of 0-45 cm

Given the fact that the field is sufficiently level, water is applied only to refill to field capacity; there was no rainfall during the experimental period and the groundwater table was far below the root zone; Ro, D, Pe, and Cf can be neglected. Accordingly the water balance equation was simplified as follows:

$$ETc = \left[\frac{\left(I + \sum_{1}^{n} (\theta_1 - \theta_2)\Delta S_i \right)}{\Delta t} \right]. \tag{6.3}$$

Where:
θ_1 = initial volumetric soil moisture content before irrigation (mm)
θ_2 = final volumetric soil moisture content after irrigation (mm)

6.2.6 Reference evapotranspiration (ETo)

The Penman-Monteith equation (Equation 6.4), which is recommended as the sole standard and accurate method of calculating ETo both by FAO and by the World Meteorological Organization (WMO) was used to compute the ETo (Allen et al., 1998).

$$ETo = \frac{\left[0.408\Delta(R_n - G) + \partial \dfrac{900}{T + 273} u_2 (e_s - e_a) \right]}{\Delta + \delta(1 + 0.34u_2)} \tag{6.4}$$

where:
ETo = reference evapotranspiration (mm)
Rn   = net radiation at the crop surface (MJm^{-2} per day)
G   = soil heat flux (MJm^{-2} per day)
T   = average air temperature (^0C)
U_2 = wind speed at 2 m height (ms^{-1})
(e_s-e_a) = vapour pressure deficit (KPa)
Δ = slope of the vapour pressure curve, ($KPa\,^0C^{-1}$)
∂ = Psychrometric constant ($KPa\,^0C^{-1}$), and 900 is the conversion factor

6.2.7 Grain yield

Grain yield was measured after maturity from pooled samples of an area of 2.0 x 1.0 m^2 in each plot (Yihun et al., 2013).

6.3 Results and discussion

6.3.1 Seasonal crop coefficient (Kc) for Teff

The overall average of Teff Kc values for the initial, development, mid-season and late-season growth stages were 0.6, 0.8, 1.2 and 0.8 respectively. The initial value of Kc started to increase after full cover of the ground, reached a maximum during the mid season stage and thereafter gradually declined. This could be explained by foliage senescence that restricted transpiration and caused a reduction in the crop coefficient.

The ETc, the ETo and the corresponding Kc values for each of the four crop growth stages (initial, development, mid-season and late-season) are summarized and presented in Table 6.1. All lysimeter experiments were conducted during the dry seasons: Lysimeter 1 (December 2010 to March 2011); Lysimeter 2 (February to May 2011); Lysimeter 3 (December 2011 to March 2012) and lysimeter 4 (February to May 2012).

6.3.2 Crop evapotranspiration

The detailed water balance components of the selected two lysimeters, the ETc, ETo and Kc values are presented in Table 6.2. The ETc for the whole growth period of Teff varied from a minimum of 299 mm to a maximum of 342 mm. The overall average was 319 mm. This average value is significantly different from the 375 mm ETc value obtained for Barley (Araya et al., 2011), but is comparable to that of Millet, which is 330 mm (Grema and Hess, 1994). Due to absence of crop water requirement data for Teff; Barley and Millet are frequently used as representatives in the design and planning of irrigation systems. These two cereals have similar shoot and root systems as Teff.

6.3.3 Crop coefficient (Kc) and days after sowing (DAS)

A multiple regression analysis with fifth-order polynomial equation was developed to determine the relationship between Days After Sowing (DAS) and Teff Kc values.

$$Kc = 7E - 8 \times 10^{-1}(DAS)^5 - 1E - 5 \times 10^{-1}(DAS)^4 + 1 \times 10^{-3}(DAS)^3$$
$$- 3.19 \times 10^{-2}(DAS)^2 + 4.88(DAS) - 2.19$$

$$(6.5)$$

$$R2 = 0.99$$

The variation in Kc as a function of DAS is shown in Figure 6.2. The coefficient of determination (r^2 = 0.99) in Equation 6.5 is as high as those obtained for similar correlations by other researchers for different crops. Kang et al. (2003) proposed a fifth-order polynomial equation for Wheat and Maize with a high coefficient of determination (r^2) of 0.96. Sepaskhah and Andam, (2001a) proposed a third-degree polynomial for Sesame crop with a coefficient of determination (r^2) of 0.85.

Table 6.1. Seasonal crop coefficients, crop evapotranspiration and reference evapotranspiration values for each growth stage of the four experimental seasons

Lysimeter	Crop Evapotranspiration in mm, Reference Evapotranspiration in mm and crop coefficient values											
	Initial stage			Developmental stage			Mid season stage			Late season stage		
	ETc	ETo	Kc	ETc	ETo	Kc	ETc	ETo	Kc	ETc	ETo	Kc
L_1	12.4	19.8	0.6	14.4	17.1	0.9	38.5	29.4	1.3	17.0	15.6	1.1
L_2	11.6	23.8	0.5	18.0	22.5	0.8	30.3	28.4	1.1	10.4	17.8	0.6
L_3	10.4	15.5	0.7	19.9	21.2	0.9	41.7	35.3	1.2	10.7	11.0	1.0
L_4	5.1	12.5	0.4	9.4	16.4	0.6	31.8	30.1	1.1	3.8	5.6	0.7
Avg	9.9	17.9	0.6	15.4	19.3	0.8	35.6	30.8	1.2	10.5	12.5	0.8

L_1, L_2, L_3 and L_4 refer to Lysimeter 1, 2, 3 and 4 in the respective four experimental seasons.

Table 6.2. Water balance components of lysimeters, ETc, ETo and Kc values at Melkassa Agricultural Research Centre

DAS	I mm	θ_i mm	Θa mm	ΔS mm	ETc mm/day	ETo mm/day	Kc
Lysimeter 2							
10	8.0	94.3	102.5	8.2	2.1	6.0	0.3
14	10.4	92.1	102.5	10.4	2.6	6.7	0.4
18	10.1	92.4	102.5	10.1	2.5	5.9	0.4
22	22.2	80.3	102.5	22.2	4.4	5.2	0.9
27	16.7	85.7	163.1	16.8	4.2	5.6	0.7
31	15.3	147.8	163.1	15.3	5.1	5.4	0.9
34	16.5	146.6	163.1	16.5	3.3	6.0	0.6
39	32.5	130.6	163.1	32.5	5.4	5.5	1.0
45	30.8	132.3	163.1	30.8	6.2	5.9	1.0
50	30.0	133.1	163.1	30.0	6.0	5.7	1.1
55	27.2	135.9	163.1	27.2	6.8	5.2	1.3
59	23.7	139.4	163.1	23.7	5.9	5.5	1.1
63	27.3	135.8	163.1	27.3	5.5	6.1	0.9
68	27.2	135.9	163.1	27.2	5.4	6.0	0.9
73	12.4	150.7	163.1	12.4	2.5	5.8	0.4
78	24.3	138.8	163.1	24.3	2.4	6.0	0.4
Lysimeter 4							
10	23.7	78.8	102.5	23.7	2.4	6.1	0.4
15	13.3	89.1	102.5	13.4	2.7	6.4	0.4
18	5.0	97.5	102.5	5.0	1.7	6.3	0.3
25	21.7	80.8	102.5	21.7	3.1	5.2	0.6
33	36.7	126.4	163.1	36.7	4.6	4.9	0.9
39	33.1	130	163.1	33.1	5.5	5.6	1.0
43	22.1	141	163.1	22.1	5.5	7.0	0.8
47	22.1	141	163.1	22.1	5.5	6.2	0.9
50	25.6	137.5	163.1	25.6	8.5	5.4	1.6
53	20.2	142.9	163.1	20.2	6.7	5.9	1.1
62	33.7	129.4	163.1	33.7	3.7	6.5	0.6
70	30.2	132.9	163.1	30.2	3.8	5.6	0.7
74	12.3	150.8	163.1	12.3	3.1	5.3	0.6

DAS refers to days after sowing, I = irrigation, θ_i and θa are the initial volumetric soil moisture content and the final volumetric soil moisture content after irrigation.

Since there was no field research conducted to determine Kc of Teff, irrigation planners and managers as well as researchers used the Kc values of Barley and Millet. As it can be deducted from Table 6.3, the Kc values of Barley and Millet are significantly different for the initial and late season stages.

6.3.4 Reference evapotranspiration calculation

Weather conditions during growing seasons

Monthly average climate variables for 2010, 2011 and 2012 seasons as well as long-term average values are presented in Table 6.4. The 2012 growing season was warmer than both 2011 and 2010 with a mean maximum temperature of 30.9 ^{0}C in 2012 and of 29.2 ^{0}C in 2011, which is higher than the mean temperature of 28.4 ^{0}C in 2010. Warmer temperatures in 2012 impacted the physiological maturity and caused large difference in the thermal time from planting to the harvest.

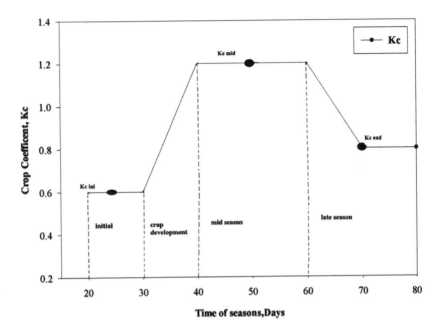

Figure 6.2. Average Kc values

Table 6.3. Crop coefficient (Kc) values of Barley, Millet and Teff grown under irrigation

Crop	Initial	Developmental	Midseason	Late season	Source
Barley	0.3	0.75	1.05-1.2	0.25-0.4	Doorenbos and kassam, 1979; Allen et al., 1998
Millet	0.3	0.75	1.05-1.2	0.3-0.4	Doorenbos et al., 1994; Allen et al., 1998
Teff	0.6	0.8	1.1	0.8	Local experimental result average values from Table 2

During the Teff growing season 27.6 mm of total precipitation was recorded in 2010, which is greater than 26.9 and 14.8 mm for 2011 and 2012 experimental years respectively. Every year much more rainfall is registered mostly in three months of the year (June, July, and August). Daily values of rainfall are shown in Figure 6.3 for the two season of Teff growth.

6.3.5 Variation of crop evapotranspiration and reference evapotranspiration

The ETc, the ETo and the corresponding Kc values for each of the four crop growth stages (initial, development, mid-season and late-season) and the four lysimeters were summarized in Table 6.1. As said, all lysimeter experiments were conducted during the dry seasons. The large variation between ETc in Lysimeters 2 and 4 are thus mainly caused by the inter-annual differences in climatic conditions and the Teff stand, which was relatively poor in the case of Lysimeter 4. The variation of reference evapotranspiration ETo and crop evapotranspiration ETc for Teff growing season is depicted in Figure 6.4.

Table 6.4. Average weather conditions during the 2010, 2011 and 2012 Teff growing seasons at Melkassa Agricultural Research Centre

Year	Month	T max (^{0}C)	Tmin (^{0}C)	T mean (^{0}C)	RH mean	Wind Speed	Rainfall (mm)	Sunshine hrs (hrs/day)
2010/2011	Nov	28.7	10.6	19.6	50.0	2.4	4.9	9.1
	Dec	27.9	5.7	16.8	47.0	2.3	3.2	8.7
	Jan	30.5	12.1	21.3	47.0	2.6	0.0	9.8
	Feb	30.4	11.9	21.1	38.0	2.8	1.5	10.1
	Mar	30.5	11.0	20.8	36.0	3.2	37.9	8.6
	Apr	32.8	10.7	21.8	38.0	2.9	45.6	8.8
	May	31.6	10.9	21.3	47.0	2.6	38.2	8.0
	June	30.9	11.5	21.3	52.0	2.7	102	8.1
2011/2012	Nov	28.0	6.7	17.3	53.0	2.3	52.2	8.7
	Dec	27.2	7.1	17.1	46.0	2.8	0.0	10.1
	Jan	28.5	11.6	20.1	50.0	2.7	0.0	9.6
	Feb	30.6	11.5	21.1	48.0	3.2	0.0	9.3
	Mar	32.6	17.1	24.9	45.0	2.4	47.5	8.6
	Apr	31.1	16.8	24.0	51.0	2.3	18.9	7.7
1991/2012	Nov	28.3	10.9	19.6	45.0	2.9	9.0	9.7
	Dec	27.6	10.9	19.4	49.0	3.2	9.3	9.5
	Jan	27.7	12.0	19.8	51.0	3.1	15.8	13.8
	Feb	29.2	13.3	21.1	49.0	3.2	24.0	8.9
	Mar	30.1	15.4	22.6	50.0	3.0	50.6	8.8
	Apr	30.1	15.4	23.4	51.0	2.8	50.9	8.4
	May	30.8	15.5	24.3	50.0	2.6	50.2	8.2
	Jun	29.8	16.3	23.9	53.0	3.2	73.7	8.2

6.4 Concluding remarks

Teff is a staple food for about 85% of the 90 million inhabitants of Ethiopia, and is currently only cultivated under rainfed agriculture with very low (1 ton/ha) yield. Due to the huge gap between demand and supply, the cost of Teff per ton has increased fivefold from 130 to 650 US$. It is hence imperative to increase the productivity under irrigated agriculture. This requires proper irrigation planning and management for which a regionally based value of the crop coefficient (Kc) is a critical parameter.

Prior to this study, as there were no properly developed Kc values for Teff, that of Barley and Millet have been used instead. As demonstrated in this study, however, the two cereal crops cannot be considered completely representative of Teff. The 0.6 Kc value for the initial and 0.8 Kc value for late season stages of Teff is double that of Barley and Millet during the same growth stages.

The overall average Teff Kc values for the initial, development, mid-season and late-season growth stages are 0.6, 0.8, 1.2 and 0.8 respectively. The Teff crop evapotranspiration (ETc) for whole growth period was measured using the water balance and found to be in the range of 299 to 342 mm. At an irrigation application of full crop water requirement the yield was three-fold of that currently harvested from rainfed farming.

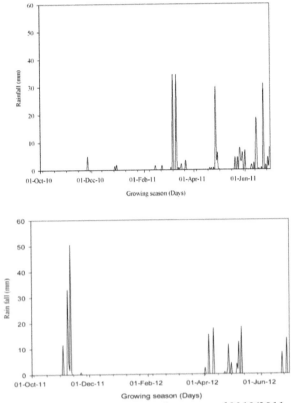

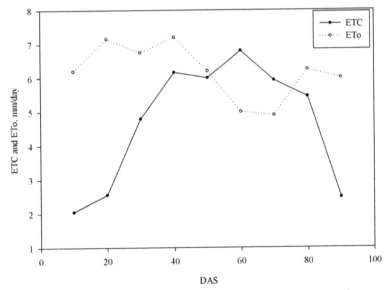

Figure 6.3. Daily rainfall for the experimental seasons of 2010/2011 and 2011/2012

Figure 6.4. Trends of average reference evapotranspiration and crop evapotranspiration of Teff for the four growing seasons

7 Application of AquaCrop in crop water productivity of Teff (*Eragrostic Tef*), a case study in the Central Rift Valley of Ethiopia

7.1 Introduction

Elaborating irrigation schedules merely on the basis of field research is expensive and time consuming, application of models is required. Several models are available to simulate yield response to water. But they are used mostly by scientists, graduate students, and advanced users. These models require an extended number of variables and input parameters not easily available for the diverse range of crops and sites. Moreover, these models present substantial complexity and are rarely used by the majority of target users such as extension personnel, water user associations, consulting engineers, irrigation and farm managers, planners and economists. Karunaratine et al. (2011) investigated suitable modelling approaches to explore the potential growth of Bambara groundnut in various agro ecological factors, but the limited nature of information from the experimental site across Sub-Sahara Africa where Bambara groundnut is grown signifies the importance of a simple model with minimum parameters.

AquaCrop, a water-driven model for use as a decision support tool in planning and scenario analysis in different seasons and locations has been developed by FAO. Although the model is relatively simple, it elaborates on the fundamental process involved in crop productivity and in the response to water deficits, both from physiological and agronomic perspective. It is designed to balance simplicity, accuracy and robustness, and is suited to address conditions where water is a key limiting factor in crop production. It requires fewer data inputs than the other models (Hsiao et al., 2009; Steduto et al., 2009). Once validated, the model is easy and needs less resource and could be useful tool in irrigation scheduling to avoid cropping risks (Tsubo et al., 2005; Soltani and Hoogenboom, 2007). Moreover, it can be used to explore and evaluate alternative management that improves water productivity and achieves more efficient water use (Bessembinder et al., 2005). The AquaCrop model simulates the variation in crop yield and biomass for the different irrigation water scenarios. Monitoring the daily water balance is required to consider the incoming and outgoing water.

Farahani et al. (2009) and Garcia Vila et al. (2009) applied the AquaCrop model for Cotton, under full and deficit irrigation regimes respectively in Syria and Spain. They showed that the key parameters, such as normalized water productivity, canopy cover and total biomass, for calibration must be tested under different climate, soil, cultivars, irrigation methods and field management conditions. Geerts et al. (2009) and Heng et al. (2009) applied the AquaCrop model to evaluate the effect of changes in the quantity of irrigation water for Quinoa, Sunflower and Maize, respectively. All researchers confirmed that the AquaCrop model can be used for scenario analysis that provides a good balance between robustness and output precision.

Teff is one of the most important and the major indigenous cereal crop of Ethiopia. It is a highly demanded cereal and has higher market prices than other cereals for both its grain and straw. The grains are highly nutritious and have high iron and protein content compared to other cereals. Moreover, Teff grain is considered as a healthy food grain for people with a problem of celiac disease as it contains no gluten (Roseberg et al., 2006). In Ethiopia more than half of the area under cereals is for Teff production. Many investigations that have been carried out indicate that Teff is adapted to environments ranging from drought-stressed to waterlogged conditions. In most parts of the country where recurrence of long periods of drought is common, it is grown as a rescue crop in case of failure of other cereals (Ketema, 1997). In spite of of this fact, with respect to Teff as a food crop in Ethiopia there has been limited research on its agronomic characteristics.

The field experimental research was carried out in Melkassa Agricultural Research Centre, located in one of the most important regions for Teff production in the country. In this region, during the rainy season the area of land covered by Teff cultivation is around 56%. The area is typically one of the most Teff grown areas, as the crop is known as a reliable and low-risk crop. The Teff growing period in most of the regions of the country is characterized by the rainy wet summer from June to August. To increase its production and to satisfy the demand, using irrigation during the dry season creates additional opportunity for smallholder as well as commercial farmers. In Tigray, northern parts of Ethiopia, calibration and validation of AquaCrop model for Teff was carried out for the rainy season using rainfall and supplementary irrigation (Araya et al., 2010). Apart from an attempt made to examine the effect of full and deficit supplemental irrigation application at flowering stage in the rainy season (Alemthay et al., 2011), there is little information on the responses to different levels of water stress at different times of the crop growth season using irrigation in the dry season.

In this chapter, the performance of application of the AquaCrop model for Teff under different regimes of irrigation (full and deficit) and different timing of irrigation application was evaluated during the dry season. The calibration of the model was done using the trial and error values of water stress coefficients for canopy expansion and stomatal closure. Moreover the measured yield and biomass with the daily monitored soil moisture data collected from the field measurements in the experimental seasons of 2010 and 2011 were used as input data for the model calibration. The validation was carried out for the independent data set of 2011/2012 experimental season. In addition the performance of simulating the response of grain yield and biomass for the different levels of irrigation was evaluated.

7.2 Field experiments

In order to evaluate the performance of the AquaCrop model, data were obtained from the field experiments conducted at the experimental fields of Melkassa Agricultural Research Centre. Cultivar of Teff (*EragrosticTef*), *Kuncho*, was selected for the evaluation of AquaCrop and its crop water productivity was assessed under the 16 different treatments. The assessments were done based on actual grain and biomass yields obtained during two irrigation seasons of the respective dry seasons: 1) November 2010 to March 2011; 2) December 2011 to April, 2012. Teff was sown on 1 December for 2010/2011 and 20 December for 2011/2012 cropping seasons. Broadcasting the Teff seed in each experimental plot was used as sowing method. The seeding rates of 25 kg/ha and 10 kg/ha were applied for both experimental seasons. Apart from water stress, specifically at the mid season stage one of the common limiting factors for high yield is lodging (Yihun et al., 2013).

The cultivar specific parameter such as emergence date was considered when 90% of the seedlings had emerged. Sowing date, flowering and duration of flowering, maximum canopy cover and maximum root depth, senescence and maturity date is observed and recorded. Senescence was assumed to be reached when the canopy started to decline. Samples of above ground dry matter production were harvested and weighted fresh and put into oven in 68 ^{0}C for 72 hrs.

The soil at the experimental site is comprised of volcanic origin formed on alluvial deposit having clay loam textural class. The soil physical properties such as field capacity are in the range of 29.3, 31.6 and 34.4%, the wilting point is 16.7, 18.4 and 21.3% (all by volume), the bulk density is 1.4, 1.5 and 1.8 g/cm^3 and the total available water (TAW) is 126, 132 and 142 mm/m for 0-30, 30-60 and 60-90 cm layers, respectively. Chemical characteristics of the soil, such as electrical conductivity (EC) values are 0.24, 0.14 and 0.20 ms/cm and pH levels vary from neutral to slightly alkaline. The value of organic matter (OM) expressed in percentage was found to be in the range of 0.98, 0.84 and 0.46. Mineral fertilizer 130 kg/ha of di-amonium nitrogen phosphate (DAP) was spread in each field just before planting. Likewise 35 kg/ha of urea was applied at tillering stage.

Daily rainfall (mm), maximum and minimum temperature (^{0}C), relative humidity (%), sunshine hours (hrs) and wind speed (m/s) were obtained from the meteorological station near (100 m) to the experimental field (Table 7.1). The long-term average growing season (rainy season) rainfall was 768 mm. The seasonal total rainfall from June to August of the study period was 555 mm and 437 mm respectively. But for the dry season irrigation experiment no effective rainfall was recorded. The daily reference evapotranspiration (ET$_O$) for the growing season was determined using the software named as ETo calculator (Raes, 2009). The average reference evapotranspiration rate during the growing season of both experimental periods was 5.4 mm/day. The daily minimum and maximum reference evapotranspiration values were 3.4 and 7.4 mm/day.

Table 7.1. Monthly average rainfall (mm), maximum and minimum temperature (^{0}C), sunshine hours and evapotranspiration at Melkassa Agricultural Research Centre for the experimental seasons of Teff from 2010 to 2012

Month	Rainfall (mm)	Temperature (^{0}C)		Sunshine (hours)	Wind speed at 2 m (m/s)	Evapotranspiration (mm/day)
		Max	Min			
November	0.6	28.3	8.7	2.4	9.1	4.9
December	0.0	27.5	6.4	2.5	9.4	4.9
January	0.0	29.1	11.6	2.8	9.8	5.3
February	1.5	29.5	12.1	3.2	10.0	5.9
March	1.9	30.2	13.4	3.1	8.6	6.0
April	1.3	31.5	14.3	2.5	7.9	5.7

Irrigation scheduling was determined based on every other day soil moisture measurement. As said irrigation water was applied manually using a watering can. The soil was generally irrigated to field capacity for those treatments receiving their full crop water requirement and satisfying their need for realization of their genetic potential. For those treatments receiving different levels of water deficit at different periods of the growth stages a specific amount of water was reduced from the full depending on their respective treatment.

7.3 Model description

AquaCrop has been widely used for research related to agricultural water management (Steduto et al., 2009). AquaCrop is a canopy-level and water driven model which simulates crop biomass and harvestable yield as constrained by water availability (full irrigation and deficit irrigation) and rainfall (Hsiao et al., 2009). It is specifically designed to predict crop productivity as a function of water stress, which is one of the most difficult relationships to accurately represent in crop modelling (Steduto et al., 2009). AquaCrop evolved from the Doorenbos and Kassam (1979) equation (7.1), for predicting yield response to water.

$$\left(1 - \frac{Ya}{Ym}\right) = Ky\left(1 - \frac{ETa}{ETc}\right) \tag{7.1}$$

where:
Ya = actual yield (corresponding to ETa) (kg/ha)
Ym = maximum theoretical yield (corresponding to ETc) (kg/ha)
ETa = actual crop evapotranspiration (mm)
ETc = potential crop evapotranspiration (mm)
Ky = proportionality factor between relative yield loss and relative evapotranspiration deficit

AquaCrop has four sub-model components: (i) the atmosphere (rainfall, temperature, reference evapotranspiration (ET_o) and carbon dioxide (CO_2) concentration); (ii) the crop (development, growth and yield); (iii) the soil (soil water balance); (iv) the management (the major agronomic practice such as fertilization and irrigation (full and deficit)). The model calculates a daily water balance that includes all the incoming and outgoing water fluxes (infiltration, runoff, deep percolation, evaporation and transpiration) and change in soil and water content. One of its advantages over the other similar models is that it separates evapotranspiration into transpiration and evaporation. Transpiration is related to canopy, which is proportional to the extent of soil cover whereas evaporation is related to the uncovered soil surface. Crops respond to water stress through four stress coefficients (reduced canopy expansion, stomata closure, canopy senescence, and change in harvest index).

The parameterization of the model involves adjusting some conservative parameters which remain fixed for a species and some site-specific parameters which are influenced by local climate, soil and management (Steduto et al., 2009). Canopy development coefficients including the maximum canopy coverage (CCx), canopy growth coefficient (CGC) and canopy decline coefficient (CDC) are key among these conservative parameters. The crop water productivity (WP*) normalized by the reference evapotranspiration (ETo), is another important conservative parameter influencing the predicted yields. The AquaCrop model distinguishes four water stress effects: on leaf expansion, stomatal conductance, canopy senescence, and harvest index (Steduto et al., 2009). Except harvest index, these effects are manifested through their individual stress coefficient, Ks, an indicator of the relative intensity of the effect. The Ks values are a function of soil water depletion and are reflected in the canopy growth of the crop. The model requires estimates of the total available water (TAW) capacity for the soil and adjusts the Ks values based on the fraction of the TAW, which has been depleted (*p*). Water stress affecting leaf expansion begins when the fraction of TAW depleted exceeds an upper threshold (p upperexp), and leaf expansion ceases when the fractional depletion exceeds a lower threshold (p lowerexp). Likewise, there are upper thresholds for water stress affecting stomatal conductance and canopy senescence, and the corresponding lower thresholds are defined by the permanent wilting point of the soil.

The advantage with AquaCrop is that it requires only a minimum input data, or the required filed data can easily be collected. Several features distinguish AquaCrop from other crop growth models achieving a new level of simplicity, robustness and accuracy that confer the model an extended extrapolation capacity to diverse locations and seasons, including future climate scenarios (Geerts et al., 2009; Raes et al., 2009; Steduto et al., 2009). The simulation of yield response under water limiting conditions will remain central in arid, semi-arid and drought-prone environments.

7.4 Model calibration

In this research AquaCrop was calibrated using the measured data set of 2010/2011 experimental season. Calibration was performed with the four irrigation treatments by first matching the ability of the fully irrigated treatment in terms of the canopy cover (CC), yield (Y) and biomass (B). After setting the growing season of the experimental year, the initial canopy cover (CCo) was estimated. Setting the sowing rate at 1000 seed mass and germination rate at the initial canopy size seedling (0.25 cm^2/plant) was estimated by the model. The initial canopy cover (CCo) 9.8% was automatically estimated by the model. The canopy growth coefficient (CGC), canopy decline coefficient (CDC) and the stress indices for water stress affecting leaf expansion and early senescence are the most important canopy cover parameters. They can be estimated by the model after some of the necessary input data of crop phenological dates, such as date to crop emergence, maximum canopy cover, senescence and maturity are entered. The water stress parameters and curve shapes were changed manually around the default value to reproduce the measured value of canopy cover. Continuous iterations of the parameters were done until satisfactory results for all the irrigation treatments in the calibrated experimental season were achieved. The detailed procedure of model calibration and the important input parameters needed for model calibration are described in Steduto et al. (2009) and Hsiao et al. (2009). The relevant input climate data used for calibration were maximum and minimum air temperature, precipitation, wind speed, sunshine hours and reference evapotranspiration (ETo). ETo was determined using the FAO Penman-Monteith equation (Allen et al., 1998), with ETo calculator software suitable for AquaCrop (Raes et al., 2009). Moreover specific data of plant growth stages were also used for model calibration.

7.5 Model validation

Validation of the AquaCrop model was done using independent data sets of the experimental season of 2011/2012. Statistical methods were used to evaluate the performance of the model. Root mean square error (RMSE) was used as one of the statistical methods.

$$RMSE = \left[\sqrt{\frac{1}{n} \sum_{i=1}^{n} (S_i - M_i)^2} \right]$$ (7.2)

where:
S_i = simulated values
M_i = measured values
N   = number of observations. Values of RMSE close to zero indicate the best fit of the model

The index of Agreement (d) is a descriptive measure and has values ranging from 0 to 1 (Willmott, 1982), which shows difference in observed and predicted mean and variance. Index of Agreement is expressed as:

$$d = 1 - \frac{\sum_{i=1}^{n}(S_i - M_i)^2}{\sum_{i=1}^{n}\left(\left|S_i - \overline{M}\right| + \left|M_i - \overline{M}\right|\right)^2} \tag{7.3}$$

Where $\overline{M}$ is the average value of the measured data; the other parameters have already been defined. The index of agreement is a descriptive measure and has a value ranging from 0 to 1. The model simulates the studied parameter better as the value approaches to 1.

7.6. Results and discussion

7.6.1 Calibration of AquaCrop model for the experimental crop Teff

The key conservative and non-conservative input parameters were used to calibrate the AquaCrop crop water productivity model. The non-conservative input soil water stress coefficients for canopy expansion, stomata closure and curve shapes were iterated to calibrate the model well. The calibrated model was tested with the independent observed experimental data set to validate the model for a series of data under different water management scenarios. The AquaCrop model simulates the observed canopy cover, yield and biomass for all levels of water application scenarios. Table 7.2 presents the crop input parameters used to simulate the target crop, Teff under the full water application scenario. The full water application scenario was used to describe crop development under non-limiting conditions in AquaCrop.

Canopy cover

The above crop parameters were used to simulate the canopy cover (CC), biomass (B) and yield (Y). The key input parameters were adjusted to obtain a good agreement between the simulated and the observed CC (Figure 7.1). The observed and simulated CC developments were fitted well for treatment receiving full irrigation and 75% of the full irrigation. Those treatments receiving higher water stress showed shorter crop cycle due to early senescence compared to the treatment receiving full irrigation. Observation from the field confirmed that treatments with severe water stress have a shorter crop cycle than treatments with no stress. As the value of sensitivity of Teff crop for aeration stress was not known, the values of Millet and Barley were used for model calibration, while these crops have the same characteristics with Teff (Grema and Hess 1994).

Treatment 1 to Treatment 4 represents treatments with 0% deficit, 25% deficit, 50% deficit and 75% deficit level respectively.

Agricultural water productivity optimization in a water scarce semi-arid region of Ethiopia

Table 7.2. Crop input parameters used in AquaCrop to simulate Teff cultivar (*Eragrostic Tef*)

Description	Value	Units	Interpretation
Planting to emergence	15	DAS	Days after sowing
Initial canopy cover (CCo)	9.38	%	Canopy size of 0.25 cm²/plant and seeding rate of 25 kg/ha
Canopy growth coefficient (CGC)	12	%/day	Increase in CC relative to existing CC per day.
Canopy decline coefficient (CDC)	15.2	%/day	Decrease in CC relative to existing CCx per day.
Maximum canopy cover (CCx)	90	%	At maximum canopy cover
Planting to maximum canopy cover	55	DAS	Days after sowing to maximum canopy cover
Planting to flowering	48	DAS	Days after sowing to flowering
Duration of flowering	15	DAS	-
Length to build up of HI	32	DAS	-
Reference harvest index (HIo)	29	%	-
Water productivity	20	g/m2	-
Planting to maximum rooting depth	60	DAS	Days after sowing to maximum root depth
Maximum rooting depth (m)	0.45	Meters	The maximum effective rooting depth
Planting to start of canopy senescence	76	DAS	Days after sowing to canopy senescence
Planting to maturity	90	DAS	Days after sowing to maturity
Shape factor for effective rooting deepening	1.5	-	-
Upper threshold for soil water stresses for canopy expansion (P upper)	0.2		fraction of Total Available Water (TAW) which has been depleted (p)
Lower threshold for soil water stresses for canopy expansion (P lower)	0.5	-	fraction of Total Available Water (TAW) which has been depleted (p)
Upper threshold for soil water stress effect on stomatal closure (P upper)	0.7		
Soil water stress on early canopy senescence (p upper)	0.79		
Aeration stress sensitivity for water logging	Sat = 10	Vol%	referred from related cereal crop (Millet)
Positive effect of soil water stress on HI before flowering	8	%	moderate effect
Soil water stress on HI during flowering (P upper7)	0.83		sensitive to water stress
Positive and negative effect of soil water stress on HI during yield formation	Small, small		small effect
Total growing period	90	DAS	growing period in the dry season

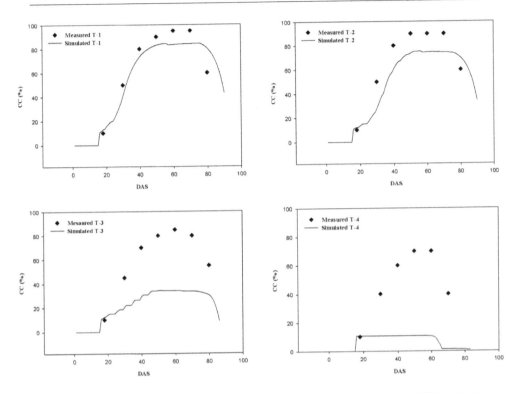

Figure 7.1. Calibration results of simulated and observed canopy cover (CC) with days after sowing (DAS) for Teff under experimental field conditions for the growing season of 2010/2011

Biomass

Simulation results of the field experimental research for biomass are shown in Figure 7.2. Those treatments receiving full irrigation (no deficit), 75% and 50% of the full irrigation water (25% and 50% deficit) reported an excellent fit with the observed Biomass. Treatment receiving full irrigation water requirement shows a very small overestimate. However, consistent under estimation was observed in those treatments having a water stress level of 75% deficit throughout the growing season indicating that the model is too severe with respect to the drought stress coefficient.

The simulated biomass and grain yield agreed well with the measured data for all treatments. For full irrigation treatment there was a strong relationship between the measured and simulated canopy and biomass with r^2 and d values of 0.87 and 0.96 for canopy and 0.97 and 0.74 for biomass respectively (Table 7.3). Even though the model simulates the grain yield better than the biomass for full irrigation application, the model under estimated the yield and the biomass under different water deficit conditions. Treatment with the stress level of 75%, the model underestimates the simulated yield and biomass compared with the measured value of yield and biomass. Similar results were recorded; as the stress level increased to the maximum the model fails its efficiency to simulate the yield and biomass (Karunaratine et al., 2011).

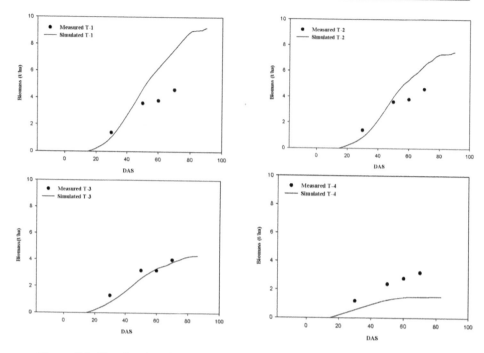

Figure 7.2. Simulated and observed biomass of Teff for the cropping season of 2010/2011 (calibration) for full irrigation application (100%) and deficit irrigation (25%, 50% and 75%) conditions

Table 7.3. Goodness of fit analysis for the simulation of canopy cover and biomass after calibration of AquaCrop

Treatment	Canopy Cover			Biomass		
	r^2	d	RMSE	r^2	d	RMSE
Treatment 1	0.87	0.96	11.0%	0.97	0.74	2.00
Treatment 2	0.84	0.91	15.7%	0.97	0.86	1.24
Treatment 3	0.93	0.57	37.2%	0.95	0.96	0.41

r^2: coefficient of determination: d: index of agreement: RMSE: root mean square error

Harvest index

The harvest index values for the different water application scenarios were obtained from the field experiment. For treatment receiving full irrigation the harvest index value obtained was 29%. The harvest index value shows a decreasing trend under water stress conditions that is 26, 23 and 20% respectively for 25, 50 and 75% deficit level. A similar trend was also reported by Farre and Faci (2009) for Maize and Sorghum for water stress at the late season stage. Moreover Karunaratne et al. (2011) reported on Bambara groundnuts for showing a decreasing trend of harvest index for severe water stress. Though the effect of soil water stresses on harvest index for Teff is small, the effect of soil water stress on different growth stage were registered and adjusted in the model. From the results of the field experiments water stress before flowering has a strong positive effect on the harvest index as a result of limited growth in the vegetative period. Water stress during yield formation had small positive and small negative effect on harvest index as a result of water stress affecting leaf expansion and inducing stomatal closure respectively. The result indicates that Teff crop can tolerate any level of water stress at the late season stage.

7.6.2 AquaCrop model validation

Validation of the model was performed with the calibrated model and using data from the independent experimental season of 2011/2012. The measured grain yield, biomass and canopy cover for all treatments were compared with the validated result of simulation. Table 7.4 presents an overview of the statistical parameters for validation of the model. For treatments receiving less amount of water stress, the model confirmed a good match between simulations and observations. As the water stress level increases, the simulated canopy cover, biomass and grain yield were underestimated. Moreover, for those treatments receiving higher stress level the observed mean is a better predictor than the model. This result was related to the shorter crop cycle of Teff due to early senescence, shorter duration of flowering and a smaller length to build up of harvest index.

Table 7.4. Statistical test of the model for grain yield and biomass

Treatment	Measured value			Validated value			Statistical Indictors		
	GY	B	CC (%)	GY	B	CC (%)	r^2	d	RMSE
Treatment 1	3.3	14	64.1	2.9	10.8	72.5	0.80	0.94	13.9
Treatment 2	3.2	11.2	61.4	2.8	10.5	69.7	0.88	0.94	13.2
Treatment 3	2.5	6.7	55.0	1.3	5.9	32.0	0.80	0.69	27.5
Treatment 4	0.6	4.5	40.8	0.5	2.4	12.1	0.39	0.45	33.6

The validation of AquaCrop model was used to simulate the yield response of Teff to variation in level of irrigation water. The data points obtained by varying the seasonal irrigation water from 100% (full crop water requirement) to 75% deficit (applying only 25% of the full crop water requirement), were used to generate the response functions needed for the optimization procedure.

7.7 Concluding remarks

Canopy, yield and biomass production under the different irrigation water application level for the experimental field conditions in the Central Rift Valley of Ethiopia were successfully simulated with the AquaCrop model. Simulation of yield and above ground biomass was satisfactory except for the severe stress levels. In addition, the AquaCrop model was able to simulate well the canopy cover of Teff under different irrigation regimes. However, the model underestimated the results of the simulation of canopy cover under the stress levels of 50% and 75%.

The harvest index value generally shows a decreasing trend under water stress conditions. Moreover, different timing of water stress has different positive and negative effects on the harvest index. Water stress before flowering has a strong positive effect on the harvest index. Water stress during yield formation had a small positive and negative effect on the harvest index.

Model validation revealed that a limited number of inputs are required to model yield response of Teff to soil water availability in Central Rift Valley. The AquaCrop model balanced between limited parameterization and good accuracy. It is therefore a powerful tool to study different water management scenarios in data scarce areas such as the sub-saharan Africa. Therefore, this model can be used to simulate the water management effects on yield and handle managements that increase water productivity.

7.8 Recommendation

Crop yield depends on many factors, including the soil fertility, amount and time of fertilizer application, and soil and water salinity. These parameters are not dealt with in AquaCrop. Therefore, some adjustments would have to be added to the model for soil and water salinity problems. Moreover, further studies and improvements could also be considered for example, the effect of soil salinity on the final yield, the effect of nutrient depletion on HI and the effect of pests, frosts and diseases on final yield have to be carried out to test the model.

8 Optimum irrigation application in irrigated Teff: moving away from exclusively rain dependent agriculture

Agriculture has retained a leading role in the recent evolution of the government of Ethiopia for sustainable development to end poverty. Ethiopia is the second most populous country in Africa (World Bank, 2013). The population of Ethiopia has increased from 24 million in 1970 to 85 million in 2012 with population growth rate of 3.2% and an expected population of 145 million by 2050 (United Nations Department of Economic and Social Affairs Population division (UNDP), 2011). Ethiopia comprises 112 Mha (million hectares) of land. Currently, high estimates show that 15 Mha of land is under cultivation. The agriculture sector is the leading sector in the economy. It accounts for 90% of the export, 85% of employment, and 55% of GDP. Though agriculture is the dominant sector, most of Ethiopia's cultivated land is under rainfed agriculture. Due to inadequate water storage to smooth and schedule the water delivery and spatial and temporal variation in rainfall, the entire agricultural cycle is disrupted whenever rain fails to come or comes too early or too late (World Bank 2006).

A large number of crops are grown in Ethiopia that includes cereals - Teff, Wheat, Barley, Corn, Sorghum and Millet -, pulses, oilseeds, vegetables, root and tubers, fruits, fibers, stimulants and Sugarcane. As Teff is only produced once in a year, the supply didn't much with its demand and the current price increased six times compared with the price in 2006. In 2010, the Teff price was still above US$ 650/ton, creating hardships for many Ethiopian families, who were forced to switch to other cereals as substitutes. Still, Teff has remained the preferred food, as evidenced by the persistently high prices over the past five years. As it is very important throughout the country, a number of biological genetic intensification ways was used to increase the genetic makeup of Teff and to increase the supply of the crop in order to maintain the price. Within the past 20 years, Teff productivity has increased by about 25 - 30%, with three-fourths of this gain attributed to the introduction of the few improved varieties that have been released and the remainder was by improved agronomic practices, mainly the application of 64 kg of di-ammonium phosphate (DAP) and 46 kg/ha of Nitrogen in the form of urea. Currently, in the experimental plots, Teff yields have reached to 3.3 ton/ha (Yihun et al., 2013). Maximum yield obtained by farmers in the farmers field is 2.5 ton/ha. However, typical farmer yields are around 1 ton/ha of grain and 5-6 ton/ha of straw used as animal feed.

To move away from exclusively rain dependent agriculture is a way to combat frequent crop failure. As the water scarcity demands the maximum use of every drop of water, there is a need to calculate the crop water productivity and economic water productivity (Bessembinder et al., 2005; Farre and Faci, 2009). The two years field experiments were done to see the effect of irrigation on Teff production. Different irrigation management and crop sowing rates were used as management scenarios to determine the optimum amount of irrigation water applied and the sowing rate needed to grow Teff. The sensitive stages of the growth season to the imposed water stress were identified.

Improving water productivity in irrigated agriculture includes increasing output per unit of water applied; reducing water loss and prioritizing water application to the most sensitive stage (Howell, 2001). Evaluation of irrigation schemes based on crop water productivity and economic water productivity per unit water applied are the most critical indicators of crop production using irrigation. However, for Teff, farmers believe the crop as a rainfed crop or they may even go to supplementary irrigation at the time rain ceases before the crop reaches to maturity. The crop water requirement, crop water productivity and economic water productivity has so far not been determined and calculated for irrigated Teff in the dry season. Therefore the objective of this study was to assess the productivity of Teff during the dry season and to determine the crop water productivity and economic water productivity for Teff under different irrigation water application scenarios of 100, 75, 50 and 25%.

8.1 Agro-ecological classification

Agro-ecological zones were classified traditionally. The traditional classes include *Bereha, Kolla, Woina Dega, Dega, Wurch and Kur*. However, at present more elaborated agro-ecological zones are established and recognized as 33, this indicates the major physical conditions that are grouped into relatively homogenous areas having similar agricultural land uses. Because of Ethiopia's location near the equator, elevation has a very strong influence on temperature and, to a lesser extent on rainfall. Generally, crop distribution is mosaic in Ethiopia. Some crops are found within several zones while others are restricted to only one or two agro-ecological zones. Even though it is one crop, in different agro-ecological zones it shows different crop water requirements and length of growing period (LGP). The LGP is defined as the number of days per year that sufficient water is available in the soil profile to support plant growth. Here, LPG is based on the number of days with a mean daily temperature above 5 ^{0}C and with available water (from precipitation or stored soil moisture) exceeding half the potential evapotranspiration. Factors that affect LPG of the crop are its variety, availability of rainfall, potential evapotranspiration, and soil moisture storage properties.

8.2 Teff production

Teff production regions in Ethiopia

Teff is one of the most important cereals grown in Ethiopia. The record of Teff yields seems to be variable across time and space: some areas record high yields in some years, but lower yields in other years. Considering the whole time period and the geographical distribution, year 2010/2011 shows higher Teff yields on a wider geographical basis. From 2006/2007 to 2010/2011 the range of yield varies from 0.6 to 1.3 ton/ha. Teff is commonly cultivated in Oromia, Amhara, Southern Nations Nationalities and Peoples' Region (SNNP) and Tigray regions. In addition, it is grown in some parts of the Benishangul Gumuz Region (Figure 8.1). The maximum yield registered at the regional level is 1.3 ton/ha (Central Statistics Authority (CSA), 2011).

Farmers and commercial growers produce Teff for both local and export markets in Ethiopia. From cereal crops Teff is a cool weather crop and grown predominantly in Ethiopian highlands at optimum altitude range of 1800 to 2200 m+MSL. Teff occupies the largest area (1.4 million hectares) and the largest cereal production. Teff, indigenous to Ethiopia, forms the staple diet of many Ethiopians and it furnishes the flour to make *Injera,* unleavened bread that is consumed in the highlands and in urban centres throughout the country. Teff is, however, a rainfed crop and produced only once in a year and harvested in *meher* (main) season.

The average mean annual rainfall in the study area is 768 mm. The main feature of seasonal rainfall in the area is inadequate in amount, poor in distribution (erratic) and intensive mainly during July and August. The rainfall follows a partial bimodal pattern with short rainy season *Belg* that extends from March-April and main rainy season *Kiremt* from July-September (Figure 8.2). The major crops grown by the farmers in the study area are Teff (*Eragrostic Tef*), Maize (*Zea mays*) and Haricot bean (*Phaseolus vulgaris*). Currently, as farmers in the area have got lucrative economic returns from Teff crop, most farmers in the area are shifting to Teff from Maize and other crops.

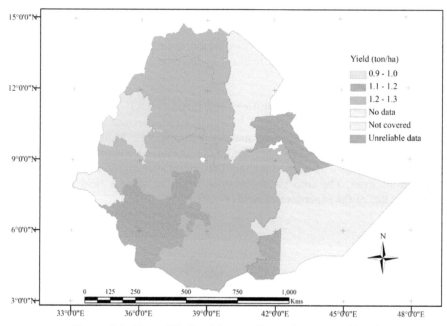

Figure 8.1. Map of Teff yield in ton/ha at the regional level

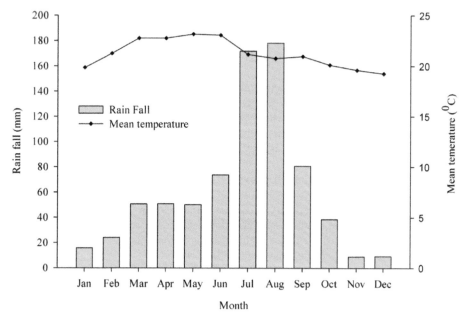

Figure 8.2. The patterns of rainfall in the area

Land preparation and tillage operations are usually done using local farm implements and animal power. Before sowing the Teff, framers of the region plough the land, two to three times using *Local Maresha* modified mouldboard plough. First tillage operation was performed on May 20 and followed by secondary tillage after 15 days. Third tillage operation was performed on July 5. Planting was done on July 10 and harvesting on October 25. Land preparation and tillage operations are usually done using local farm implements and animal power. Considering the long series rainfall data of the study area and discussions with the surrounding farmers and researchers in the experimental station, the area experiences erratic events and sometimes remains dry almost all the year. This results in a reasonable decrease of the quantity and the quality of the harvestable yield.

8.3 Gap in crop yield in research stations and farmers field

Farmers do lack inputs and improved management practices. Moreover, poor research extension linkage and limitation of promotion of recommended management practices is a problem. Thus, there is a need to improve input availability and affordability to farmers in all areas. The crop yields obtained from the farmers field is substantially smaller than that of the experimental field (Figure 8.3).

Figure 8.3. Comparison of farmer's field with the experimental research field in the rainy season

In the experimental fields, a maximum of 3.3 and 12 ton/ha were obtained for grain and straw yields. However, typical farmer grain and straw yields are still very low to 1 and 6 ton/ha for grain and straw yield respectively. Contrary to common belief that Teff yields have reached a ceiling, Teff grain and straw yields can be doubled and even tripled by field management and drastically reducing plant population.

8.4 Field management practices improving Teff production and water productivity

Despite its importance, the productivity of Teff is much lower than of other cereals. The national average yield is about 1 ton/ha, compared to 2.8 ton/ha for Maize and 1.9 ton/ha for Wheat. For Teff the major yield-limiting factor is its susceptibility to lodging (falling over), especially during the grain filling period. As a result, the quality and quantity of the grain is severely reduced and results in low grain and straw yield. As compared to other cereal crops Teff is known to be a resistant crop both at the time of planting (on the field) and after harvest (low post harvest loss). From the experimental field observation one of the main reasons related with yield loss is lodging, lodging is falling of the stem of the Teff at the time of flowering.

8.4.1 Method of irrigation water application

The method of irrigation water application and frequency of irrigation are equally important to deliver the required amount of irrigation water to attain optimum production.

In both experimental years, experimental field trials were made for sprinkler irrigation and furrow irrigation. From the analysis of field observations, applying irrigation water using sprinkler irrigation from top aggravates the falling of the stem of the Teff plant. While irrigating Teff from the bottom using the furrows maintains the Teff stand and decreases the potential of the very thin stem of the Teff to fall over from the pressure of the irrigation water (Figures 8.4 and 8.5).

Figure 8.4. Method of application of irrigation water for the experimental Teff field

8.4.2 Rate of sowing

A sowing rate of 25 kg/ha (used by most farmers and research stations) and a smaller seeding rate of 10 kg/ha was used to evaluate the impact of varying the sowing rate on the grain yield and straw. From the analysis of the experimental field observations decreasing the sowing rate to 10 kg/ha increases the tillering potential of the crop. Moreover, compared to the sowing rate of 25 kg/ha, the 10 kg/ha field shows strong stem and is less inclined to lodging (Figure 8.6).

Figure 8.5. Furrow irrigation for Teff planted in rows

Figure 8.6. Higher tillering potential of the 10 kg/ha sowing rate

8.5 Economic water productivity of irrigated Teff

Economic water productivity (EWP) is expressed in gross income in US$ per gross water supplied in m^3. EWP was computed based on the following information obtained at the study site: the size of irrigable area, the price of water applied, and the income gained from the sale of grain (main product) and the straw (by-product) considering the average seasonal local market price (US$) as shown in Equation 8.1. The gross income is the product of the average price of Teff per kg for the season and the average grain yield per given irrigable area plus the product of the price of Teff straw per kg of the season and the average straw yields per given irrigable area. Local market prices were considered, because almost all the products are consumed locally under the present conditions.

Increase in crop production per unit of water doesn't necessarily result into an increase in the farmer's income because of the non-linearity of crop yield with the price of products. Table 8.1 shows the economic water productivity and crop water productivity of Teff for different irrigation application scenarios.

Given a certain amount of water, scheduled and measured irrigation application depth for four levels of irrigation water application scenarios of 100, 75, 50 and 25% of the crop water requirement. Teff EWP was calculated for the experimental seasons of the growing cycle.

The mean potential grain and straw yield of Teff cultivar for the experimental seasons in the study site was 3.3 and 14 ton/ha respectively. The mean seasonal price of grain and straw yield of Teff around Central Awash Rift Valley was 700 and 80 US$/ton respectively.

$$EWP = \frac{Gross\,income\,(US\$)}{Irrigation\ water\,(m^3)} \qquad\qquad (8.1)$$

$$GI = (PTG * YLD_g) + (PTS * YLD_s) \qquad\qquad (8.2)$$

where:
EWP   = economic water productivity in US$/m^3
GI   = gross income in US$
PTG   = price of Teff grain per ton of the season in US$/m^3
YLD$_g$ = average grain yield per given irrigable area ton/ha
PTS   = price of Teff straw per ton of the season in US$/m^3
YLD$_s$ = average straw yield per given irrigable area ton/ha

The economic water productivity calculated for the 2010/2011 experimental season under 100, 75, 50 and 25% irrigation application scenario were 1.17, 1.26, 1.21 and 1.16 US$/m^3 respectively (Table 8.1). The same trials were made for the experimental season of 2011/2012, the economic water productivity was found to be 0.71, 0.87, 1.11 and 0.56 US$/m^3 for the four scenarios of water management respectively. As the experiments took place in the dry season, no rainfall and only irrigation was concerned. Higher EWP was found to be 1.26 and 1.11 respectively for the experimental seasons of 2010/2011 and 2011/2012. Therefore, saving the water with 25% and 50% to increase the area irrigated result in increasing Teff production.

Even though, Teff is one of the high value crops, under a given water supply; vegetables have high economic water productivity. It seems wise to give priority to crops with higher EWP and CWP so that cost of investment could be quickly regained as well as more products per unit water applied could be obtained.

Under a water scarce environment, EWP and CWP could be improved through crop choice, reducing water losses through optimizing the field application efficiency, optimum irrigation application and marketability. From the field survey and field observations most farmers of the region prefer to produce onion and tomato than cereals. Even though they know that Teff has relatively higher local market price and social acceptance than other cereals, they still choose to produce vegetables for their day to day survival.

Table 8.1. Estimated EWP and CWP for the four levels of water application scenarios

Experimental Season	Parameters	Water application level scenario			
		100%	75%	50%	25%
2010/2011	Yield (ton/ha)	3.12	2.45	1.51	0.69
	Straw (ton/ha)	10.7	9.34	6.52	3.40
	Irrigation water (mm)	281	211	140	70
	EWP (US$/m^3)	1.17	1.26	1.21	1.16
	CWP (kg/m^3)	1.11	1.16	1.08	0.98
2011/2012	Yield (ton/ha)	3.3	3.18	2.52	0.66
	Straw (ton/ha)	14.0	11.2	6.7	4.5
	Irrigation water (mm)	380	294	225	156
	EWP (US$/m^3)	0.71	0.87	1.11	0.56
	CWP (kg/m^3)	0.87	1.08	1.12	0.42

100, 75, 50 and 25% are different water application scenarios in reference to the full crop water requirement of Teff (100%), EWP and CWP are economic water productivity and crop water productivity.

8.6 Conclusions

Irrigation substantially increased the Teff grain yield. It is the way to combat frequent failure of Teff yield due to erratic nature of rainfall in the region. Despite of its importance, the productivity of Teff is much lower than of other cereals. The national average is 1 ton/ha. The crop yields obtained from the farmers field is substantially smaller than that of the experimental field. The main reasons for low Teff yield in farmer's fields are lack inputs and improved management practices. Moreover, poor research extension linkage and limitation of promotion of recommended management practices is a problem. Thus, there is a need to improve input availability and affordability to farmers in all areas. Currently, as farmers in the region have got lucrative economic returns from Teff, most farmers shift to production of Teff during the dry season using irrigation.

For higher crop water productivity, knowing the crop water requirement helps in optimum production and in turn results in saving of water to increase the irrigated area.

The values of EWP and CWP for Teff under different irrigation management scenarios were developed. Higher EWP and CWP values of 1.26 and 1.16 were obtained by minimizing irrigation water application by 25%. Therefore, the EWP and CWP could be used as a tool for decision making in growing crops under limited water conditions. For appropriate evaluation of economic water productivity and crop water productivity of Teff, the special criteria of Teff such as social acceptance, production cost, health benefit, storage and export values and other crucial factors should be considered.

9 Conclusions and Recommendations

9.1 Conclusions

Due to the rapid growth in population, the pressure on water resources is increasing. In the near future less water will be available for agricultural production due to competition with the domestic and industrial sectors, while at the same time food production must be increased to feed the growing population. It is inevitable that the production per unit water consumed, the water productivity must be increased to meet this challenge.

This thesis illustrates the strategy for sustainable food production in drought prone areas like the Central Rift Valley of Ethiopia. Improving the productivity of land and water is the only option for sustainable food production. Increasing food production to feed the fast growing world population particularly in developing countries relies on two options: either increase the arable land or improve the productivity of the existing cultivated land. In the drought prone and highly degraded areas, rainfed cereal production is restricted by low and highly variable rainfall with common occurrence of dry spells during the cropping season on one hand, and low levels of soil fertility on the other hand. Improving agricultural water productivity is the greatest potential to increase food production in irrigated agriculture to feed the fast growing population in water scarce environments. In the subsequent paragraphs a summary of important conclusions drawn from each application is given.

Teff (*Eragrostic Tef*) is an endemic cereal crop and staple food for about 85% of the 90 million inhabitants of Ethiopia. It is highly preferred by the farmers not only in view of its adaptability to a wide range of environmental conditions but also for its high market value of both grain and crop residue (straw). However, low productivity is often reported mainly due to water stress. The production of Teff according to the farmers' sowing practice is strongly limited by water stress as farmers use rainfall for Teff production. Even though the time of sowing is at the rainy season, the low and highly variable rainfall with the common occurrence of dry spells during the cropping season cannot satisfy the crop water requirement need. For example, in the study area, maturity is far from reached at the end of the rainy season. Planning for optimal sowing time is therefore crucial. Moreover; the water demand of the crop exceeds the rainfall supply in the region.

Irrigation substantially increased the Teff grain yield. The yield continues to increase as the amount of water applied to increase up to a certain level. For a higher economic water productivity, knowing the crop water requirement of the crops helps in optimum production and in turn results in saving of water to increase the irrigated area. As the crop water requirement of Teff is not exactly determined, many agronomists in the region assume that Teff water requirement is proportional to Barley and Millet. In contrast to the subjective assumption, the Teff crop evapotranspiration (ETc) for whole growth period was measured using the water balance and found to be in the range of 299 to 342 mm. Prior to this study, as there were no properly developed Kc values for Teff, that of Barley and Millet have been used instead. The 0.6 Kc value for the initial and 0.8 for late season stages of Teff is double that of Barley and Millet during the same growth stages. As demonstrated in this study, however, the two cereal crops cannot be considered completely representative of Teff.

Considering this context, a research project was designed with two types of field experiments to observe the effect of irrigation on Teff production. The field experiments were carried out in the dry seasons of 2010/2011 and 2011/2012. Different irrigation management and crop sowing rates were used as management scenarios to determine the optimum amount of irrigation water applied and the sowing rate needed to grow Teff. The sensitive stages of the growth season to the imposed water stress were identified. A selected combination of depth of irrigation water application (amount) and growth stage (time) of Teff (*Eragrostic Tef*) was used as experimental design in order to determine the optimum water application depth at specific growth stages that result in optimum crop water productivity (CWP). This research investigated the sensitivity of each growing stage to drought stress in detail. Four different levels of irrigation water supply were applied, full crop water requirement 0% deficit (ETc), 25% deficit (applying 75% of crop water requirement), 50% deficit (applying 50% of crop water requirement) and 75% deficit (applying 25% crop water requirement). The phenological cycle was divided into phases, which are considered to be most relevant from the viewpoint of their response to irrigation, i.e. initial stage (P1), development stage (P2), mid season stage (P3) and late season stage (P4). From the result of the crop responses to the given irrigation water application the following major conclusion could be drawn - maximum grain yields of 3.12 and 3.3 ton/ha were obtained for the seeding rate of 25 kg/ha. For the seeding rate of 10 kg/ha, maximum yields of 2.91 and 3.05 ton/ha respectively were obtained from the treatment which received the optimum crop water requirement (satisfying the full crop water requirement) during the experimental seasons. This is three fold the yield farmers currently harvest from rainfed agriculture.

In case of treatment 2 (75% of ETc irrigation application i.e. 25% deficit) the yield values were reduced to 2.45 and 2.27 ton/ha respectively. Much more significantly lower yields of 0.69 and 0.45 ton/ha were obtained for treatment 4 (25% of ETc irrigation application i.e. 75% deficit) throughout the whole growth stage for both 25 kg/ha and 10 kg/ha seeding rates. Moreover, 75, 50 and 25% irrigation water reduction throughout the whole growth stage decreased the Teff yield by 77.9, 51.6 and 21.5% respectively. Treatment 7 (75% deficit) III, Treatment 11 (50% deficit) III and Treatment 15 (25% deficit) III which were conducted under adequate watering conditions throughout the first two periods of the growing season, and followed by a period of stress at the mid season stage with 75, 50 and 25% deficit water application resulted in the second, the third, and the fourth lowest yield respectively by 30, 23 and 23% for the seeding rate of 25 kg/ha. This yield reduction is significant compared with stressing the crop during late season stage having a reduction of 11%.

The crop was very sensitive to water stress during the mid season stage. Even when it was only subject to 25% deficit, its yield decreased by about 1 ton/ha. Stressing the crop either by one-half or three-quarters at the mid season stage, results in lower yields next to stressing the crop throughout the growth season.

The yield and water productivity differences are insignificant between full irrigation and 25% deficit irrigation distributed throughout the growth period at seeding rates of 25 kg/ha and 10 kg/ha. Thus, when water is scarce and irrigable land is relatively abundant as is the case in Ethiopia, adopting the 25% water deficit irrigation with 10 kg/ha seeding rate is recommended. A maximum water deficit of 50% during the late season stage has an insignificant impact on Teff yield and water productivity. The values of economic water productivity (EWP) and crop water productivity (CWP) for Teff under different irrigation management scenarios were developed. Higher EWP and CWP values of 1.26 and 1.16 were obtained by minimizing irrigation water application by 25%. The obtained values varied with the years and ranged from 1.12 to 1.16 kg/m^3 and 1.08 to 1.31 kg/m^3 for the seeding rate of 25 kg/ha and 10 kg/ha respectively. Therefore, the EWP and CWP could be used as a tool for decision making in growing crops under limited water conditions. For appropriate evaluation of economic water productivity and crop water productivity of Teff, the special criteria of Teff such as social acceptance, production cost, health benefit, storage and export values and other crucial factors need to be considered.

In addition, also at Melkassa Agricultural Research Centre, two years experiments with four lysimeters were conducted with irrigated Teff during the dry seasons to determine the specific Kc values of Teff for each growth stage. The overall average Teff Kc values for the initial, development, mid-season and late-season growth stages were 0.6, 0.8, 1.2 and 0.8 respectively.

In this study the crop water productivity model AquaCrop developed by FAO was used as a tool to test different field management strategies for improving productivity of Teff under the local environmental conditions in the Central Rift Valley of Ethiopia. The field experiments, conducted in 2010/2011 and 2011/2012 were used to gain insight in the response of Teff to different water stress levels and to calibrate and validate the AquaCrop model for Teff under water stress conditions. Additionally, data collected during field experiments conducted in the same experimental seasons from the four lysimeter experiments were also used as additional inputs for the calibration process.

The simulation analysis with AquaCrop revealed that the model was adequately able to simulate the yield of Teff in response to water stress under various environmental conditions. This was confirmed by accurate estimation of the soil water balance and canopy development that determine the crop transpiration, biomass and yield production. With the calibrated and validated model, irrigation water management strategies' including the different water stress levels throughout the whole growth season were evaluated for two years. The model estimates yield by relating crop transpiration with biomass and yield production and allows the users to simulate yield under various conditions. Canopy, yield and biomass production for the different irrigation water management under the experimental field conditions in the Central Rift Valley of Ethiopia were successfully simulated by AquaCrop.

The harvest index value shows generally a decreasing trend under water stress conditions that is 26, 23 and 20% respectively for 25, 50 and 75% deficit level. Moreover, different timing of water stress has different positive and negative effect on the harvest index. The effect of soil water stress on different growth stages were registered and adjusted in the model. Water stress before flowering has a strong positive effect on the harvest index as a result of limited growth in the vegetative period. Water stress during yield formation had small positive and negative effect on the harvest index as a result of water stress affecting leaf expansion and inducing stomatal closure respectively. This indicates, Teff crop can tolerate any level of water stress at the late season stage.

Model validation was undertaken for the different water stress levels. For treatments receiving less amount of water stress such as no stress (0% deficit), 25% deficit, 50% deficit level; the model confirmed that there is a good agreement between simulations and observations with coefficient of determination (r^2) = 0.80, index of agreement (d) = 0.94 and root mean square error (RMSE) = 13.9. As the water stress level increases with the 75% deficit level, the simulated canopy cover, biomass and grain yield were underestimated with r^2 = 0.39, d = 0.45 and RMSE = 33.6. Moreover, for those treatments receiving higher stress level the observed mean is a better predictor than the model. Model validation revealed that a limited number of inputs are required to model yield response of Teff to soil water availability in Central Rift Valley. The AquaCrop model balanced between limited parameterization and good accuracy, and it is therefore a powerful tool to study different water management scenarios. Therefore, this model can be used to simulate the water management effects on yield and handle managements that increase water productivity.

9.2. Recommendations

The importance of agricultural water productivity concept for food security where water resources are rapidly being exhausted has been discussed in the thesis. This study is limited to one crop, Teff. Most of the analysis was done in reference to a limited scope of the agro-climatic constraints and alternative measures to improve the soil water availability for growing Teff in a drought prone semi-arid part of the Northern Ethiopia. In future research, we recommend a nationwide characterization of the climate and crop water management alternative using the new approach developed in this thesis. Furthermore, future research is recommended on crop varietal difference for most cereal crops to see the effect on water use efficiency and water productivities.

Use of the validated AquaCrop model has the advantages of simplicity, low cost and low data requirement, and its application is water driven, which is important for the exploring yield under various water availability conditions in water scarce environments. However, the model would have to be refined for different agronomic conditions such as plant population, cultivar difference, soil and nutrient conditions as described in chapter 6 of this thesis.

Crop yield depends on many factors, including the soil fertility, amount and time of fertilizer application, and soil and water salinity. These parameters are not dealt with in AquaCrop. Therefore, further studies and improvements should also be considered for example, the effect of soil salinity on the final yield, the effect of nutrient depletion on Harvest Index and the effect of pests, frosts and diseases on final yield have to be carried out to test the model.

Other recommendations:
- it is important to further analyze the effect of different water stress levels to the nutrient and fertilizer utilization of Teff;
- this study considered the effect of application of different levels of irrigation water amount on yield of Teff keeping all the other components constant. However, future studies would also have to consider the quality of irrigation water. Moreover, field scale experiments on the analysis of soil type and quality of soil help to better estimate the agricultural production.

10 References

Abou, K. and Abdrabbo, A. 2009. Macro management of deficit-irrigated peanut with sprinkler irrigation. Agricultural Water Management 96(10): p. 1409-1420.

Alemthay, T., Raes, D., and Geerts, S., 2012. Unravelling crop water productivity of Teff (Eragrostic Tef (Zucc.) Trotter) through AquaCrop in Northern Ethiopia: Experimental Agriculture, v. 48, no. 2, p. 222-237.

Ali, M. and Talukder, M., 2008. Increasing water productivity in crop production - a synthesis: Agricultural water management, v. 95, no. 11, p. 1201-1213.

Allen, R.G., Pereira, L.S., Raes, D. and Smith, M., 1998. Crop evapotranspiration-Guidelines for computing crop water requirements-FAO Irrigation and drainage paper 56: FAO, Rome, v. 300, p. 6541.

Antony, E. and Singandhupe, R., 2004. Impact of drip and surface irrigation on growth, yield and WUE of capsicum: Agricultural water management, v. 65, no. 2, p. 121-132.

Araya, A., Keesstra, S., and Stroosnijder, L., 2010. Simulating yield response to water of Teff (Eragrostic Tef) with FAO's AquaCrop model: Field Crops Research, v. 116, no. 1-2, p. 196-204.

Araya, A., Stroosnijder, L., Girmay, G. and Keesstra, S.D., 2011. Crop coefficient, yield response to water stress and water productivity of Teff (Eragrostic Tef (Zucc.): Agricultural water management, v. 98, no. 5, p. 775-783.

Assefa, K., Merker, A. and Tefera, H., 2003. Inter simple sequence repeat (ISSR) analysis of genetic diversity in Tef [Eragrostic Tef (Zucc.) Trotter]: Hereditas, v. 139, no. 3, p. 174-183.

Bennie A.P., Hensley, M., 2001. Maximizing precipitation utilization in dry land agriculture in South Africa Journal of Hydrololgy 241: 125-139.

Bessembinder, J., Leffelaar, P., Dhindwal, A. and Ponsioen, T., 2005. Which crop and which drop, and the scope for improvement of water productivity: Agricultural water management, v. 73, no. 2, p. 113-130.

Brisson, N. and Casals, M.L., 2005. Leaf dynamics and crop water status throughout the growing cycle of durum wheat crops grown in two contrasted water budget conditions: Agronomy for Sustainable Development, v. 25, no. 1, p. 151-158.

Bossie, M., Tilahun, K. and Hordofa, T., 2009. Crop coefficient and evaptranspiration of onion at Awash Melkassa, Central Rift Valley of Ethiopia: Irrigation and Drainage Systems, v. 23, no. 1, p. 1-10.

Burman, R., Wright, J., Nixon, P. and Hill, R., 1980a. Irrigation management - water requiements and water balance. In: Irrigation, challenges of the 80's1980a, American Society of Agricultural Engineer, St Joseph, p. 141-153.

Central Statistics Authority (CSA), 2007. Population and Housing Census.Addis Ababa,Ethiopia.

Central Statistics Authority (CSA), 2010. Statistical Data of Ethiopia, Addis Ababa, Ethiopia.

Central Statistics Authority (CSA), 2011. Atlas of agricultural statistics 2006/07 - 2010/2011 Addis Ababa, Ethiopia.

Central Statistics Authority (CSA), 2012. Statistical Data of Ethiopia, Addis Ababa, Ethiopia

Charles, R., 2002. Ground Water Science. An Imprint of Elsevier Science, 525 B Street, Suite 1900, San Diegi, California 92101-4495, USA, ISBN No. 0-12-257855-4. p. 4.

Cosgrove, W.J., Rijsberman, F.R., 2000. World Water Vision: Making Water Everybody's Business Earthscan Publications London.

Deng, X.P., Shan, L., Zhang, H. and Turner, N.C., 2006. Improving agricultural water use efficiency in arid and semi-arid areas of China: Agricultural water management, v. 80, no. 1, p. 23-40.

Dinka, T.M., 2007. Application of the Morgan, Morgan Finney Model in Adulala Mariyam Watershed, Ethiopia.

Doorenbos, J. and Pruitt, W., 1977. Crop water requirement. Rome: FAO, 1977. 144p: Irrigation and Drainage Paper, v. 24.

Doorenbos, J.K., 1979. Guidelines for predicting crop water requirements -Yield response to water: Irrigation and Drainage, p. 193.

Engida, M., 2000. A desertification convention based on moisture zone of Ethiopia. Journal of NaturalResource 2000(1). p. 1-9.

Evert, S., Staggenborg, S. and Olson, B., 2009. Soil Temperature and Planting Depth Effects on Tef Emergence: Journal of Agronomy and Crop Science, v. 195, no. 3, p. 232-236.

Famine Early Warning Systems Network Ethiopia (FEWSNET), 2012. Ethiopia price bulletine, Addis Ababa,Ethiopia.

Farahani, H.J., Izzi, G. and Oweis, T.Y., 2009. Parameterization and evaluation of the AquaCrop Model for full and deficit irrigated cotton. Journal of Agronomy 101:469-476.

Food and Agriculture Organization of the United Nations (FAO), 2002. Deficit irrigation practice. Water Reports 22 Rome, Italy.

Food and Agriculture Organization of the United Nations (FAO), 2005. AQUASTAT, Helping to build a world without hunger. , 2002. Water 2002. Reports 29. Rome, Italy.

Food and Agriculture Organization of the United Nations (FAO), 2010. Agriculture Organization. 2003: State of the World's Forests. FAO, Rome, Italy.

Farre, I. and Faci, J.M., 2009. Deficit irrigation in maize for reducing agricultural water use in a Mediterranean environment: Agricultural water management, v. 96, no. 3, p. 383-394.

Fereres, E., Soriano, M.A. 2007. Deficit irrigation for reducing agricultural water use. Special issue on Integrated approaches to sustain and improve plant production under drought stresses. Journal of Experimental Botany, 58: 147-159.

García-Vila, E.F., Mateos, L., Orgaz, F. and Steduto, P., 2009. Deficit irrigation optimization of cotton with AquaCrop. Agricultural Water Management 101: p. 477-487.

Geerts, S. and Raes, D., 2009. Deficit irrigation as an on-farm strategy to maximize crop water productivity in dry areas: Agricultural water management, v. 96, no. 9, p. 1275-1284.

Grema, A. and Hess, T., 1994. Water balance and water use of pearl Millet-cowpea intercrops in north east Nigeria: Agricultural water management, v. 26, no. 3, p. 169-185.

Habtegebrial, K. and Singh, B., 2006. Effects of timing of nitrogen and sulphur fertilizers on yield, nitrogen, and sulphur contents of Teff (Eragrostic Tef (Zucc.) Trotter): Nutrient Cycling in Agroecosystems, v. 75, no. 1, p. 213-222.

Heng, L.K., Hsiao, T., Evert, S., Howell, T. and Steduto, P., 2009. Validating the FAO AquaCrop Model for irrigated and water deficient field maize. Journal of Agronomy 101:488-498.

Howell, T., Piccinni, G., Ko, J., Marek, T., 2009. Determination of growth-stage-specific crop coefficients of Maize and Sorghum: Agricultural water management, v. 96, no. 12, p. 1698-1704.

Hsiao, T.C., Heng, L., Steduto, P., Rojas-Lara, B., Raes, D. and Fereres, E., 2009. AquaCrop The FAO Crop Model to Simulate Yield Response to Water: III. Parameterization and Testing for Maize. Journal of Agronomy 101:488-459. http://www.mower.gov.et/wresurfacewatertblclimate.php

Hussain, I. and Hanjra, M.A., 2004. Irrigation and poverty alleviation: review of the empirical evidence: Irrigation and Drainage, v. 53, no. 1, p. 1-15.

Intergovernmental Panel on Climate Change (IPCC), 2007. Climate change: Impacts, adaptation and vulnerability. In: M.L. Parry,O.F. Canziani, J.P. Palutikof, P.J. van der Linden and C.E. Hanson (Eds.), Contribution of working group II to the forth assessment, report of the intergovernmental panel on climate change. Cambridge Univesity Press. Cambridge. United Kingdom.

Irmak, S., Mutiibwa, D., Irmak, A., Arkebauer, T., Weiss, A., Martin, D. and Eisenhauer, D., 2008. On the scaling up leaf stomatal resistance to canopy resistance using photosynthetic photon flux density: Agricultural and Forest Meteorology, v. 148, no. 6, p. 1034-1044.

Itanna, F., 2005. Sulphur distribution in five Ethiopian Rift Valley soils under humid and semi-arid climate. Journal of Arid Environmentt.62: 597-612.

International Water Management Institute (IWMI), 2000. World water supply and demand: 1995-2025. Colombo, Sri Lanka. Available at http://www.iwmi.cgiar.org/.

Jensen, M.E., 1974. Consumptive use of water and irrigation water requirements: Irrigation and Drainage Division American Society of Civil Engineering, New York.

Jensen, M.E., Burman R.D. and Allen R.G., 1990. Evapotranspiration and irrigation water requirement. American society of civil engineers:Newyork.

Kang, S., Zhang, L., Xiaotao, H., Huanie, C. and Binjie, G. 2002. Effects of limited irrigation on yield and water use efficiency of winter wheat in the Loess Plateau of China. Agricultural Water Management 55(3): 203-216.

Kang, S., Gu, B., Du, T. and Zhang, J., 2003. Crop coefficient and ratio of transpiration to evapotranspiration of winter wheat and maize in a semi-humid region: Agricultural water management, v. 59, no. 3, p. 239-254.

Karunaratne, A.S.,Azam-Ali, S. N.,Al-Shareef, I., Sesay, A., Jorgensen, S. T. and Crout, N.M.J., (2010). Modelling the canopy development of bambara groundnut. Agricultural and Forest Meteorology 150:1007-1015.

Karunaratne, A.S.,Azam-Ali, S. N.,Izzi, G and Steduto, P., 2011. Caliberation and Validation of FAO-AquaCrop model for irrigated and water deficit Bambara groundnut: Expermental Agriculture 10.1017.

Kashyap, P.S. and Panda, R., 2001. Evaluation of evapotranspiration estimation methods and development of crop-coefficients for potato crop in a sub-humid region: Agricultural water management, v. 50, no. 1, p. 9-25.

Katerji, N., Mastrorilli, M. and Cherni, H.E., 2010. Effects of corn deficit irrigation and soil properties on water use efficiency. A 25-year analysis of a Mediterranean environment using the STICS model: European Journal of Agronomy, v. 32, no. 2, p. 177-185.

Ketema, S., 1997. Tef (Eragrostic Tef (Zucc.)Trotter). Promoting the conservation and use of underutilized and Neglected Crops. Institute of Plant Genetics and Crop Plant Research, Gatersleben/International Plant Genetic Resources Institute, Rome, Italy.

Kijne, J.W., Barker R., Molden D., 2003. Water Productivity in Agriculture: Limits and Opportunities for Improvement. CAB International, Wallingford, United Kingdom.

Kijne, J.W., Tuong, T.P., Bennett, J., Bouman, B. and Oweis, T., 2003b. Ensuring food security via improvement in crop water prodcutivity. In: challenge programm on water and food: background papers to the full proposal. The challenge programm on water and food consortium, Colombo, Sir Lanka.

Kipkorir, E.C., Raes, D. and Massawe, B. 2002. Seasonal water production functions and yield response factors for maize and onion in Perkerra, Kenya. Agricultural Water Management 56(3): 229-240.

Kirda, C., 2002. Deficit irrigation scheduling based on plant growth stages showing water stress tolerance: Irrigation Science, p. 3-10.

Kloss, S., Pushpalatha, R., Kamoyo, K.J. and Schütze, N., 2012. Evaluation of crop models for simulating and optimizing deficit irrigation systems in arid and semi-arid countries under climate variability: Water resources management, v. 26, no. 4, p. 997-1014.

Majnooni-Heris, A., Sadraddini, A.A., Nazemi, A.H., Shakiba, M.R., Neyshaburi, M.R. and Tuzel, I.H., 2012. Determination of single and dual crop coefficients and ratio of transpiration to evapotranspiration for canola: Scholars Research Library.

Makombe, G., Kelemework, D. and Aredo, D., 2007. A comparative analysis of rainfed and irrigated agricultural production in Ethiopia. Irrigation and Drainage Systems 21(1): p. 35-44.

Mamo, G. 2006. Using seasonal climate outlook to advice on sorghum production in the Central Rift Valley of Ethiopia. PhD. Thesis, University of Free State, PhD. Thesis, University of Free State, Bloemfontein.

Meinzen-Dick, R.S., Rosegrant, M.W. (Eds.). 2001. Overcoming water scarcity and quality constraints. Washington DC, USA: IFPRI. p. 28. (2020 Vision for Food, Agriculture and the Environment, focus 9).

Mengistu, D.K., 2009, The influence of soil water deficit imposed during various developmental phases on physiological processes of Tef (Eragrostic Tef): Agriculture, Ecosystems & Environment, v. 132, no. 3, p. 283-289.

Ministry of Agriculture (MOA). 2000. Agro-Ecological Zones of Ethiopia, Natural Resource Management and Regulatory Department, Addis Ababa, Kirda, Ethiopia.

Ministry of Water Resources (MOWR). 2001. Water and development Ministry of Water Resource Magazine December 2001 No. 20.

Molden, D.J., Rijsberman, F. 2001. Assuring water for food and environmental security. In: Consultative Group on International Agricultural Research (CGIAR) Mid-Term Meeting, Durban, South Africa, 26 May.

Molden, D., Murray-Rust, H., Sakthivadivel, R. and Makin, I., 2003. A water productivity Frame work for Understanding and Action. Water productivity in agriculture: Limits and opportunities for Improvement, ed. J.W. Kijne. CABI, ISBN No.0851996698, Wallingford, United Kingdom:

Molden, D.J., Oweis, T.Y., Steduto, P., Kijne, J.W., Hanjra, A.H. and Bindraban, P.S., 2007. Pathways for increasing agricu;tural water productivity. In: Molden, D.(ed) Water for food, water for life: a comprehensive assesement of water management in agriculture. Earthscan, London and International Water Management Institute, Colombo, Sirlanka, p. 279-310.

Molle, F., Wester, P. and Hirsch, P., 2010. River basin closure: processes, implications and responses. Agricultural Water Management, v. 97, no.4, p. 569-577.

Moutennet, P., 2002. Yield response factors of field crops to deficit irrigation. In: FAO Water Reports, No 22, p. 11-15.

Payero, J.O., Tarkalson, D.D., Irmak, S., Davison, D. and Petersen, J.L., 2008. Effect of irrigation amounts applied with subsurface drip irrigation on corn evapotranspiration, yield, water use efficiency, and dry matter production in a semi-arid climate: Agricultural water management, v. 95, no. 8, p. 895-908.

Perry, C., 2007. Efficient irrigation; inefficient communication; flawed recommendations: Irrigation and Drainage, v. 56, no. 4, p. 367-378.

Piccinni, G., Ko, J., Marek, T. and Howell, T., 2009. Determination of growth-stage-specific crop coefficients of maize and sorghum: Agricultural water management, v. 96, no. 12, p. 1698-1704.

Raes, D., 2009. ETo Calculator: a software program to calculate evapotranspiration from a reference surface. FAO Land Water Division: Digital Media Service, no. 36.

Rijsberman, F., 2001. Can the Consultative Group on International Agricultural Research (CGIAR) solve the world water crisis? Paper presented at the CGIAR Mid-Term Meeting 2001 in Durban, South Africa, on 26 May.

Rijsberman, F.R., 2006. Water scarcity: Fact or fiction? Agricultural Water Management 80(1-3): p. 5-22.

Roseberg, R.J., Norberg, S., Smith, J., Charlton, B., Rykbost, K. and Shock, C., 2006. Yield and quality of Teff forage as a function of varying rates of applied irrigation and nitrogen: Research in the Klamath Basin 2005 Annual Report. OSU-AES Special Report, v. 1069, p. 119-136.

Schultz, B., Tardieu, H., Vidal, A., 2009. Role of water management for global food production and poverty alleviation. Irrigation and Drainage 58 (S1): S3-S21.

Sepaskhah, A.R. and Andam, M., 2001a. Crop coefficient of sesame in a semi-arid region of I.R. Iran: Agricultural water management, v. 49, no. 1, p. 51-63.

Sepaskhah, A.R. and Andam, M., 2001b. Crop coefficient of sesame in a semi-arid region of IR Iran: Agricultural water management, v. 49, no. 1, p. 51-63.

Soltani, A., and Hoogenboom, G., 2007. Assessing crop management options with crop simulation models based on generated weather data. Field Crops Research 103, p. 198-207.

Spaenij-Dekking, L., Kooy-Winkelaar, Y. and Koning, F., 2005. The Ethiopian cereal Tef in celiac disease: New England Journal of Medicine, v. 353, no. 16, p. 1748-1749.

Stadler, S.J., 2005. Ardity indexes. In J.E. Oliver (Ed.), Encyclopedia of world climatology. Heidelberg: Springer, p. 89-94.

Steduto, P., Hsiao, T. C., Raes, D., and Fereres, E., 2009, AquaCrop - The FAO crop model to simulate yield response to water: I. Concepts and underlying principles: Agronomy Journal, v. 101, no. 3, p. 426-437.

Stewart, J., Misra, R., Pruitt, W., And Hagan, R., 1975. Irrigating Corn And Grain Sorghum With A Deficient Water Supply.

Tefera, B., Ayele, G., Atnafe, Y., Jabbar, M. and Dubale, P., 2002. Nature and causes of land degradation in the: Oromiya Region: A review.

Tiwari, K., Singh, A. and Mal, P., 2003. Effect of drip irrigation on yield of cabbage (Brassica oleracea L. var. capitata) under mulch and non-mulch conditions: Agricultural Water Management, v. 58, no. 1, p. 19-28.

Tsubo, M., Walker, S., Ogindo, H. O., 2005. A simulation model of cereal-legume intercropping systems for semi-arid regions I. Model develeopment. Field Crop Research. 93, p. 10-22.

United Nations Environment Program (UNEP).1992. World atlas of desertification. Editorial Commentary by N.J.Middleton and D.S.G. Thomas London: Edward Amold.

United Nations Educational, Scientific and Cultural Organization (UNESCO). 1979. Map of the world distribution of arid regions: Map at scale 1:25,000,000 with explanatory note. MAB Technical Notes 7, UNESCO, Paris.

United Nations Department of Economic and Social Affairs Population division (UNDP), 2011. World population prospects: The 2010 Revision, http://esa.un.org/unpd/wpp/index.htm on April 2011 by United Nations.

United State central Intelligence Authority, 2009. World Factbook Library of Congress Classification, Retrieved July 3, 2012, from http://archive.org/ details/theciaworldfactb35829gut.

United State Library of Congress, 2005. Ethiopia, Country Studies Handbook, Retrieved July 17, 2006, from http:// country studies.us/Ethiopia.

Vorosmarty, C.J., Green, P., Salisbury, J. and Lammers, R.B., 2000. Global water resources: vulnerability from climate change and population growth. Science 289, p. 284-288.

Welderu fael, W.A. 2006. Quantifying Rainfall-Runoff Relationships on selected Benchmark Ecotopes in Ethiopia: A Primary step in water Harvesting Research. PhD, Thesis, University of the Free State, Bloemfontein, South Africa.

Willmott, C.J., 1982. Some Comments on the eevaluation of model performance. Bulletine of the American Meteorological Society 63, p. 1309-1313.

World Bank, 2006. Ethiopia: managing water resource to maximize sustainable growth. A world bank Water Resource Assistance Strategy for Ethiopia, Washington DC, USA.

World Bank, 2013. Ethiopia Overview Available at:http://www.worldbank.org/en/country/ethiopia/overview (up dated April 2013, accessed May 2013).

Yihun, Y.M., Schultz, B., Haile, A.M and Erkossa, T., 2011. Optimizing Teff water productivity in water stressed region of Ethiopia. ICID 21st, Tehran, Iran October 2011.

Yihun, Y. M., Haile, A. M., Schultz, B. and Erkossa, T., 2013. Crop Water Productivity of Irrigated Teff in a Water Stressed Region: Water resources management, p. 1-11.

Zhang, J. and Yang, J., 2004. Improving harvest index is an effective way to increase crop water use efficiency, p. 21-25.

Zhang, Y., Kendy, E., Qiang, Y., Changming, L., Yanjun, S. and Hongyong, S. 2004. Effect of soil water deficit on evapotranspiration, crop yield, and water use efficiency in the North China Plain. Agricultural Water Management 64(2), p. 107-122.

Annex I. Samenvatting

Verwacht wordt dat de wereldbevolking zal groeien van 7 miljard nu tot 9 miljard in 2050. De levensstandaard in de opkomende landen (bijna 75% van de wereldbevolking) stijgt snel. In de minst ontwikkelde landen is er over het algemeen een snelle groei van de bevolking. De bevolkingsgroei in combinatie met de stijging van de levensstandaard vereisen een substantiële toename van de voedselproductie om duurzame voedselzekerheid te garanderen.

De landbouwsector wordt vandaag de dag in feite geconfronteerd met een complexe reeks uitdagingen: het produceren van meer voedsel van een betere kwaliteit met minder water per eenheid van product, het produceren van duurzame voeding en het verminderen van ondervoeding, het toepassen van schone technologieën die zorgen voor ecologische duurzaamheid, het hoofd bieden aan mogelijke gevolgen van klimaatverandering en om op een productieve manier bij te dragen aan de lokale en nationale economie, van producent tot consument. Daarom is het verbeteren van waterbeheer in de landbouw van cruciaal belang voor het behoud van de mondiale voedselzekerheid en de bestrijding van armoede in het landelijke gebied. Geïrrigeerd land beslaat nu ongeveer 20% van het wereldwijde landbouwareaal en zorgt voor 55% van de wereld voedselproductie. Wereldwijd vertegenwoordigt irrigatie 70% van de onttrokken hoeveelheid water. Het verminderen van het gebruik van irrigatiewater door het verhogen van de productiviteit is cruciaal om aan de steeds toenemende vraag naar water voor de landbouw te voldoen. Zo wordt geïrrigeerde landbouw geconfronteerd met de druk om zijn aandeel in het waterverbruik te verminderen en tegelijkertijd zorg te dragen voor de productie van voldoende voedsel en vezels voor een groeiende bevolking en andere behoeften.

De snelle toename van de bevolking vereist een adequaat beheer van land en water in Ethiopië. De bevolking van Ethiopië is gestegen van 24 miljoen in 1970 tot 85 miljoen in 2012 met een bevolkingsgroei van 3,2% en een verwachte bevolking van 145 miljoen in 2050. De verdeling van de Ethiopische bevolking is over het algemeen gerelateerd aan de agro-ecologische kenmerken en gunstige topografische omstandigheden. De economie van het land is sterk afhankelijk van de landbouw, die op haar beurt afhankelijk is van de beschikbaarheid van seizoensgebonden regenval. Landbouw is ondenkbaar zonder land en water. Landbouw kan alleen effectief zijn wanneer het wordt voorzien van voldoende water op het juiste moment. Ongeveer 85% van de Ethiopische bevolking is afhankelijk van regenafhankelijke landbouw. Onvoldoende seizoensgebonden regenval kan leiden tot ernstige voedseltekorten, waardoor het sociale en economische leven van de mensen kan worden gedestabiliseerd. De meeste regio's van Ethiopië hebben te lijden van onvoldoende en onbetrouwbare regenval. Als gevolg van ernstige bodemerosie in de hooglanden, de grond in deze regio is ondiep, is er een zeer laag vochthoudend vermogen en een laag gehalte aan organische stof. De ongunstige omstandigheden met betrekking tot klimaat en bodem leiden voor het grootste deel van het jaar tot het voorkomen van vochttekorten in de bodem wat resulteert in verlies aan plantaardige productie. In de meeste regio's is er alleen regen in enkele maanden van het groeiseizoen en deze is meestal kort en intens wat resulteert in een hoge afvoer, die ongebruikt optreedt. De intensiteit van terugkerende droogte beïnvloed het levensonderhoud van agrarische gemeenschappen en de economie als geheel. De situatie wordt nog verergerd door de afnemende kwaliteit van water en bodem. Dientengevolge is voedselonzekerheid nog steeds het grootste probleem en een grote zorg voor het land. Daarom is het noodzakelijk om grote droge, semi-droge en sub-humide gebieden met een onregelmatige regenval verdeling te gaan irrigeren.

Irrigatie heeft een veelzijdige rol in het bijdragen tot voedselzekerheid, zelfvoorziening, productie en export van levensmiddelen. Schaarse watervoorraden en toenemende concurrentie om water zal de beschikbaarheid ervan voor irrigatie verminderen. Tegelijkertijd, is er de noodzaak om aan de groeiende vraag naar voedsel te voldoen door verhoogde productie van gewassen met minder water. Het bereiken van een grotere efficiëntie in het watergebruik is een primaire uitdaging voor de nabije toekomst en zal de toepassing van technieken en praktijken die leiden tot een meer accurate levering van water aan gewassen omvatten.

Omdat de bevolkingsdichtheid in de hooglanden van Ethiopië bleef stijgen, zijn bovendien meer en meer marginale gronden in cultuur gebracht, wat uiteindelijk resulteerde in een ernstige verslechtering van de agro-ecologische hulpbronnen en een dalende landbouwproductie. Dien ten gevolge is de uitbreiding van de bevolking in de richting van de uitgestrekte laagland gebieden (droog en semi-droog) toegenomen. Helaas zijn er in gebieden waar het water vanwege periodieke droogten schaars is onzekerheden in de watervoorziening die ernstige gevolgen kunnen hebben voor de watervoorraden en hun duurzaamheid voor een optimale productie kunnen bedreigen. Verhoging van de productiviteit van het water in de landbouw door een hogere productie onder een efficiënter gebruik van water is een belangrijke strategie voor de aanpak van waterschaarste. Naast vergroting van het geïrrigeerde gebied, hebben verhogen van de intensiteit en gewasopbrengsten geholpen om de voedselproductie per hoofd van de bevolking te stabiliseren, hoewel de bevolking en de consumptie per hoofd van de bevolking aanzienlijk zijn gegroeid. Dit vraagt om de toepassing van geschikte innovatieve technologieën voor een betere en duurzame agrarische productie en productiviteit.

Omdat wateraanvoer doorgaans beperkt is, is de doelstelling van boeren om het netto inkomen per eenheid gebruikt water te maximaliseren. Aangezien er een redelijke prijsstijging is van de landbouwproductie, is veel aandacht nodig om de water productiviteit (WP) te verhogen. Onlangs is de nadruk gelegd op het concept van water productiviteit, hier gedefinieerd als de productiviteit van water bij een bepaald gebruik in termen van kwantiteit en kwaliteit van het onttrokken water. In dit verband verwijst de term water productiviteit naar de hoeveelheid of de waarde van het product ten opzichte van het volume of de waarde van het onttrokken water. Afhankelijk van hoe de termen in de teller en de noemer worden uitgedrukt, kan de productiviteit van water worden uitgedrukt in algemene fysische en economische termen. Fysische productiviteit wordt gedefinieerd als de hoeveelheid product gedeeld door de onttrokken hoeveelheid water (kg/m^3). Economische productiviteit van water wordt gedefinieerd als de waarde per eenheid water of de netto huidige waarde van het product gedeeld door de netto huidige waarde van de onttrokken hoeveelheid water. Het verhogen van de productiviteit van water in de landbouw zal een vitale rol spelen bij het verlichten van de concurrentie om het schaarse water, het voorkomen van aantasting van het milieu en het verstrekken van voedselzekerheid. Molden en Rijsberman geven een eenvoudig bewijs voor de bovenstaande stelling: door het verbouwen van meer voedsel met minder water, zal meer water beschikbaar zijn voor andere natuurlijke en menselijke toepassingen. Het is daarom noodzakelijk om de bestaande toewijzing en verdeling van watersystemen te onderzoeken, evenals gewas patronen en intensiteiten, om de huidige beperkingen die verbetering behoeven te identificeren. Daarom was dit onderzoek gericht op het ontwikkelen van een strategie om de beperkte hoeveelheid beschikbaar water zo efficiënt mogelijk te gebruiken en om innovatieve technologische benaderingen en methoden voor waterbeheer te helpen identificeren voor een optimale landbouwproductie en water productiviteit.

Een groot aantal gewassen worden geteeld in Ethiopië. Dit betreft granen (Teff, Tarwe, Gerst, Maïs, Sorghum en Gierst), peulvruchten, oliehoudende zaden, groenten, wortels en knollen, fruit, vezels, versterkende middelen en Suikerriet. Van de granen staat Teff bekend als een regen afhankelijk gewas dat slechts eenmaal per jaar geoogst wordt. Omdat het maar een keer per jaar wordt geoogst, voorziet de levering niet in de vraag en is de huidige prijs zes keer zo hoog in vergelijking met de prijs in 2006. In 2010 was de Teff prijs nog steeds boven US$ 780/ton, wat zorgde voor ontberingen voor veel Ethiopische gezinnen, die hierdoor gedwongen werden naar andere granen als vervangers om te schakelen. Toch is Teff het favoriete voedsel gebleven, zoals blijkt uit de aanhoudend hoge prijzen in de afgelopen vijf jaar. Omdat het in het hele land erg belangrijk is, is een aantal biologisch genetische manieren van intensivering toegepast om de genetische samenstelling van Teff te verhogen en de levering van het gewas te vergroten om de prijs te handhaven. In de afgelopen 20 jaar is de productiviteit van Teff met ongeveer 25-30% toegenomen, waarbij driekwart van deze winst is toe te schrijven aan de invoering van de weinige verbeterde rassen die zijn vrijgegeven en de rest was door verbeterde agronomische praktijken, vooral de toepassing van 64 kg di-ammoniumfosfaat (DAP) en 46 kg/ha stikstof in de vorm van ureum. Momenteel worden in de experimentele velden Teff opbrengsten tot 3,3 ton/ha bereikt. De maximale opbrengst die door boeren op hun akkers wordt verkregen bedraagt 2,5 ton/ha. Typische opbrengsten voor de boeren zijn echter ongeveer 1 ton/ha graan en 5-6 ton/ha stro dat wordt gebruikt als diervoer.

Teff (*Eragrostic Tef*) is het belangrijkste inheemse graan gewas van Ethiopië, waar het vandaan komt en is gediversifieerd. Het is zeer gewild en basisvoedsel voor meer dan 85% van de 85 miljoen inwoners. Teff heeft hogere marktprijzen dan de andere granen voor zowel de korrel, als het stro. De boeren verdienen meer voor de teelt van Teff dan voor de teelt van andere granen. Boeren en commerciële kwekers produceren Teff voor lokale en exportmarkten. Meer dan de helft van het graan areaal is voor Teff productie. Teff wordt niet aangetast door kevers, waardoor het geringere verliezen bij opslag na de oogst heeft en geen chemicaliën voor controle op ziekten nodig zijn. De vele onderzoeken die zijn uitgevoerd geven aan dat Teff is aangepast aan omgevingen variërend van droogte tot drassige omstandigheden.

Teff meel wordt voornamelijk gebruikt om een gefermenteerde, zure soort deegachtig, plat brood genaamd Injera te maken. Teff wordt ook gegeten als pap of gebruikt als een ingrediënt voor zelf gebrouwen alcoholische dranken. Het heeft ook een hoog ijzergehalte en een hoog potentieel als export gewas naar de Verenigde Staten en Europese landen, want het bevat geen gluten en wordt beschouwd als een gezond voedsel graan. Ongeveer een miljoen Amerikanen lijden aan coeliakie (gluten gevoeligheid) en Teff kan een niche invullen voor het voldoen aan dieetvoorschriften. Er worden serieuze pogingen gedaan om de teelt uit te breiden in Europa, vooral in Nederland, en in de Verenigde Staten.

Eeuwenlang en tot 5 jaar geleden is Teff alleen verbouwd als een regen afhankelijk gewas en werd het slechts een keer in een jaar geoogst met maximale opbrengst van 1 ton/ha. Aangezien de prijs van Teff in het afgelopen decennium is verhoogd van 2000 Birr (ongeveer US$ 130) per ton tot 10.000 Birr (US$ 650) per ton, is het verhogen van de productie van het gewas met inbegrip van het toepassen van irrigatie belangrijk geworden.

Afstappen van uitsluitend regen afhankelijke landbouw is een manier om de frequente misoogsten te bestrijden. Twee soorten veldproeven zijn uitgevoerd om het effect van irrigatie op Teff productie analyseren. De veld experimenten zijn in de droge seizoenen van 2010/2011 en 2011/2012 uitgevoerd. Verschillend irrigatiebeheer en zaad hoeveelheden zijn gebruikt als beheer scenario's om de optimale hoeveelheid irrigatiewater en de hoeveelheid zaad te bepalen die nodig zijn om Teff te laten groeien. De gevoelige stadia van het groeiseizoen op de opgelegde water tekorten zijn geïdentificeerd. Tevens zijn, eveneens bij MARC, twee jaar experimenten met vier lysimeters uitgevoerd met geïrrigeerde Teff gedurende de droge seizoenen.

Het gedetailleerde onderzoek om de reactie van de opbrengst van Teff, biomassa en water-gewas productiviteit voor verschillende scenario's van irrigatiewater toepassing te bepalen is als volgt uitgevoerd. In de opzet van de experimenten zijn combinaties van levering van irrigatiewater (hoeveelheid) en groeistadium (tijd) van Teff (*Eragrostic Tef*) gebruikt om de optimale watergift bij specifieke groeistadia te bepalen, die resulteerde in een optimale water-gewas productiviteit (CWP). Bij dit onderzoek is de gevoeligheid van elke groeifase op droogte stress in detail onderzocht. Vier verschillende niveaus van de levering van irrigatiewater zijn toegepast: volledige waterbehoefte van het gewas 0% tekort (ETc), 25% tekort (het toepassen van 75% van de waterbehoefte van het gewas), 50% tekort (het toepassen van 50% van de waterbehoefte van het gewas) en 75% tekort (toepassen van 25% van de waterbehoefte van het gewas). De fenologische cyclus werd opgedeeld in fasen die vanuit het oogpunt van hun reactie op toediening van irrigatie water als het meest relevant worden beschouwd, dat wil zeggen initiële fase (P1), ontwikkelingsfase (P2), midden groei fase (P3) en rijping fase (P4). Een vier bij vier combinatie van zestien behandelingen met drie herhalingen werd in de veld experimenten toegepast met een totaal van achtenveertig proeven. Elke set van deze 48 proeven werd getest met zaad hoeveelheden van 10 en 25 kg/ha. Derhalve bedroeg het totale aantal experimentele velden op het Melkassa Centrum voor Landbouwkundig Onderzoek (MARC) voor het eerste experimentele seizoen 96. Elk experimenteel veld had een oppervlakte van 8 m^2 (2,0 x 4,0 m^2). De individuele velden werden door middel van een kleine wal van elkaar gescheiden.

Het Teff ras, ter plaatse genaamd, *Kuncho* was geselecteerd en de water-gewas productiviteit werd bepaald op basis van de 16 verschillende behandelingen. *Kuncho* werd gekozen omdat in vergelijking met andere lokale variëteiten die zijn vrijgegeven door MARC, het de meest favoriete is onder de lokale bevolking en het een hoge marktwaarde in het land en de regio heeft. De bepalingen zijn gedaan op basis van werkelijke graan en biomassa opbrengsten gedurende twee irrigatie seizoenen: 1) november 2010 tot maart 2011 en 2) december 2011 tot april 2012. De gegevens van de belangrijkste fenologische stadia, zoals zaaidatum, datum van 90% opkomst, 50% bloei, de duur van de bloei, volgroeiing en rijping zijn geregistreerd. De plant hoogte is met een interval van tien dagen gemeten bij hetzelfde monster in elke veld. Bovengrondse biomassa waarnemingen zijn eveneens om de tien dagen gedaan op basis van ongestoorde monster gebiedjes van 1 m^2. De bovengrondse biomassa werd in een droogoven gedurende 48 uur bij 60 ^{0}C gedroogd en vervolgens gewogen. Graanopbrengsten werden na de oogst gemeten op basis van samenstelde monsters van een gebied van 2,0 x 3,0 m^2 per veld. Het gewas werd handmatig geoogst. Het graan en het totale biomassa verse gewicht werden gewogen op de dag van de oogst en vervolgens gedroogd en gewogen op een nauwkeurige weegschaal. Bodemvocht werd regelmatig gemeten met behulp van een neutronen sonde. Irrigatie water werd toegediend op basis van de verschillende condities voor tekort aan water. Bij een volledige irrigatie (geen tekort), werd het bodemvocht constant op veldcapaciteit gehouden. De neutronen sonde werd gekalibreerd met de gravimetrische methode waarbij voor het zaaien bodemmonsters zijn verzameld op diepten van 15, 30 en 45 cm en vervolgens elke 5-7 dagen tot aan rijping. Om the komen tot volume basis is het gemeten bodemvocht op gewicht basis vermenigvuldigd met de dichtheid.

Het bleek dat het verminderen van de algemeen aanbevolen zaai hoeveelheid van 25 kg/ha tot 10 kg/ha de sterkte van de Teff steel vergrootte en een grotere weerstand van de stam bood tegen knakken door de hoge pluimen bij rijping. Bovendien verhoogde de kleinere zaai hoeveelheid het potentieel voor uitstoeling. Bij de zaai hoeveelheid van 10 kg/ha werd bij de behandeling op basis van de optimale waterbehoefte van het gewas voor de experimentele seizoenen een maximale opbrengst van verkregen van respectievelijk 2,91 en 3,05 ton/ha. Bij het zaaien van 25 kg/ha werd een graan opbrengst van 3,12 en 3,3 ton/ha verkregen. Dit is het drievoud van de opbrengst die boeren momenteel oogsten bij regenafhankelijke teelt. In het algemeen waren de patronen van respons op irrigatie behandeling, ongeacht de seizoenen, vergelijkbaar en toonden een significante en positieve reactie. Wanneer 100% van ETc werd toegepast (0% tekort - behandeling T1), was de graan opbrengst voor Teff tijdens 2010/2011 bij het zaaien van respectievelijk 25 kg/ha en 10 kg/ha 3.12 en 2.91 ton/ha. In het geval van behandeling T2 (75% van ETc irrigatie toediening, dat wil zeggen tekort 25%) gingen de opbrengsten omlaag tot respectievelijk 2,45 en 2,27 ton/ha.

Significant veel lagere opbrengsten van 0,69 ton/ha en 0,45 ton/ha werden verkregen voor behandelingen met toediening van 75% te weinig irrigatie water (T4) gedurende de hele groei voor zowel 25 als 10 kg/ha zaaien. Bovendien verminderde toediening van 75, 50 en 25% irrigatiewater gedurende de gehele groeifase de opbrengst Teff met respectievelijk 77,9%, 51,6% en 21,5%. Behandelingen die werden uitgevoerd met voldoende irrigatiewater in de eerste twee perioden van het groeiseizoen, gevolgd door een periode van water tekort in de midden groei fase met T7 (75% tekort), T11 (50% tekort) en T15 (25% tekort) resulteerde in de tweede, de derde en de vierde laagste opbrengst respectievelijk met 30%, 23% en 23% bij het zaaien van 25 kg/ha. Deze opbrengst reductie is significant in vergelijking met water tekort voor het gewas tijdens het late groei stadium met een reductie van 11%. Water tekort voor het gewas, hetzij met de helft of driekwart in het midden van het groei seizoen, resulteerde in lagere opbrengsten naast het watertekort voor het gewas gedurende het hele groeiseizoen. Een maximaal water tekort van 50% tijdens de rijping fase had een verwaarloosbare invloed op de Teff opbrengst en water productiviteit. Aangezien de prijs van Teff zaad met een factor tien steeg kan het verminderen van de zaaidichtheid zaad besparen. Hogere waarden voor de water-gewas productiviteit zijn verkregen met de behandeling op basis van 75% ETc. De verkregen waarden varieerden met de jaren en bedroegen 1,12-1,16 kg/m^3 en 1,08-1,31 kg/m^3 voor het zaaien van respectievelijk 25 kg/ha en 10 kg/ha. De opbrengst en water-gewas productiviteit verschillen waren niet significant tussen volledige irrigatie en irrigatie op basis van 25% tekort over de hele groeiperiode. Dus, wanneer water schaars is en irrigeerbaar land relatief overvloedig, zoals het geval is in Ethiopië, wordt toepassing van irrigatie op basis van 25% tekort met zaaien van 10 kg/ha aanbevolen.

Ondanks het belang van Teff als voedselgewas in Ethiopië, is er slechts beperkt onderzoek naar de agronomische en fysiologische reacties op water en andere fysieke stress gedaan. Bovendien is in droge en semi -droge gebieden, zoals de Centrale Awash Rift Valley (het specifieke studiegebied), het water schaars en is het verhogen van de water-gewas productiviteit noodzakelijk voor duurzame voedsel productie en water veiligheid. Echter, nauwkeurige bepaling van de gewas coëfficiënt (Kc), de verhouding tussen de gewas verdamping (ETc) en referentie gewas verdamping (ETo), is een belangrijke factor voor effectieve irrigatie planning en beheer.

De gewas coëfficiënt (Kc) voor Teff is nauwkeurig bepaald met de lysimeters. De lysimeters waren van het niet-weegbare type, twee ervan waren vierkant in doorsnede met 2 m lang, 2 m breed en 2 m diep en de andere twee waren rechthoekig met 2 m lang, 1 m breed en 2 m diep. Ze waren gemaakt van gewapend beton en de binnenkant was bekleed met een plastic folie om lekkage of laterale in- en uitstroom van water te voorkomen. Hat aanwezige bodemvocht, wat een van de belangrijke invoergegevens voor de berekening van gewas verdamping (ETc) is, werd om de dag bepaald met een neutronen sonde op basis van de wortel diepte van Teff. Voor een grotere nauwkeurigheid werd de neutronen sonde met de gravimetrische methode op basis van de vaststelling van natte en droge punten gekalibreerd voor het bodemtype van de experimentele velden, teneinde een breed scala aan vochtgehalten te vinden en het mogelijk te maken deze bereiken met de sonde te bepalen. Irrigatie werd toegepast op het gewas als ongeveer 55% van het beschikbare bodemvocht in de effectieve wortelzone was onttrokken zoals aanbevolen door Allen et al. (1998). Het tekort werd omgezet in volume basis en handmatig met een gieter met een bekend volume gesuppleerd.

De actuele Teff ETc werd gemeten op basis van de waterbalans van de vier lysimeters, de gemiddelde ETc voor de hele groeiperiode van het gewas werd bepaald op 320 mm. De ETo werd berekend op basis van dagelijkse weergegevens voor het onderzoeksgebied met behulp van software genaamd ETo rekenmachine. De verkregen Kc waarden waren 0.6, 0.8, 1.1 en 0.8 voor respectievelijk de initiële fase, ontwikkelingsfase, midden groei fase en rijping fase. Vergeleken met de Kc waarden van Gerst en Gierst, bekend en gebruikt als een substituut voor Teff, waren de waarden voor Kc van Gerst en Gierst voor de verschillende groeistadia respectievelijk als volgt bepaald 0.3, 0.75, 1,05-1,2 en 0,25-0,4. Het grote verschil in de waarde van Kc in de initiële en rijping fasen toont het belang van de regionale vaststelling van het gewas zelf, anders dan gebruik te maken van de Kc waarde van soortgelijke gewassen.

Uitwerking irrigatie schema's op basis van louter veldonderzoek is lastig en tijdrovend, toepassing van modellen is vereist. Verschillende gewas modellen zijn beschikbaar om gewas opbrengst bij toediening van water te simuleren. Ze worden echter vooral gebruikt door wetenschappers, studenten en gevorderde gebruikers omdat deze modellen een groot aantal variabelen en invoerparameters vereisen, die niet gemakkelijk beschikbaar zijn voor de diversiteit aan gewassen en locaties. Bovendien zijn deze modellen van een aanzienlijke complexiteit en worden ze zelden gebruikt door de meerderheid van de beoogde gebruikers, zoals landbouw voorlichters, verenigingen van watergebruikers, raadgevende ingenieurs, managers van irrigatie systemen en boerderijen, planners en economen.

AquaCrop, is een op water gebaseerd model dat is ontwikkeld door de Voedsel en Landbouw Organisatie van de Verenigde Naties (FAO) voor gebruik als een beslissing ondersteunend hulpmiddel bij het plannen en scenario analyse voor verschillende seizoenen en locaties. Hoewel het model relatief eenvoudig is en minder gegevensinvoer vereist dan de meeste andere modellen, legt het de nadruk op het fundamentele proces dat van toepassing is op de productiviteit van water bij de teelt van gewassen en in de reactie op water tekorten, zowel in fysiologische als in agronomisch opzicht. Het is ontworpen om eenvoud, nauwkeurigheid en robuustheid evenwichtig te benaderen en is geschikt voor omstandigheden waarbij water een belangrijke beperkende factor is voor gewasproductie. De belangrijkste parameters zijn genormaliseerde productiviteit van water, oogst index, bedekkingsgraad, opbrengst en biomassa. Het model kan de variatie in gewasopbrengst en biomassa voor verschillende scenario's van toepassing van irrigatiewater simuleren. Het AquaCrop model is bijvoorbeeld toegepast om het effect van veranderingen in de hoeveelheid irrigatiewater voor respectievelijk quinoa, tarwe, zonnebloemen en maïs, te evalueren. Die onderzoekers toonden aan dat het een model voor scenario-analyse is dat een goede balans oplevert tussen robuustheid en nauwkeurigheid van het resultaat.

In het huidige onderzoek werd AquaCrop gekalibreerd voor Teff in de droge seizoenen van 2010 en 2011. Onafhankelijke datasets zijn gebruikt ter validatie van het model. Model validatie werd uitgevoerd voor de verschillende niveaus van water tekort. Voor behandelingen die waren gebaseerd op een geringere hoeveelheid water tekort, bevestigde het model dat er een goede overeenkomst is tussen de simulaties en waarnemingen met correlatie coëfficiënt (r^2) = 0,80, index van overeenkomst (d) = 0.94 en wortel uit het gemiddelde kwadraat van de fouten (RMSE) = 13.9 waarden. Als het water tekort niveau stijgt, werden de gesimuleerde bedekking graad, biomassa en graan opbrengst onderschat met r^2 = 0,39, d = 0,45 en RMSE = 33,6 waarden. Bovendien was voor de behandelingen op basis van een groter watertekort het waargenomen gemiddelde een betere voorspeller dan het model. Bij de model validatie bleek dat een beperkt aantal invoergegevens nodig was om de reactie van de opbrengst van Teff op bodemvocht gehalten in de Centrale Rift Vallei te bepalen. Het AquaCrop model vertoonde een goed evenwicht tussen beperkte invoer parameters en goede nauwkeurigheid, en het is dus een doeltreffend instrument om verschillende waterbeheer scenario's te bestuderen. Daarom kan dit model worden gebruikt om effecten van waterbeheer op de opbrengst van Teff te simuleren en richtingen aan te geven die de water productiviteit verhogen.

Al met al heeft het onderzoek het potentieel en de beperkingen van het combineren van experimenteel veldwerk met modelleren voor het optimaliseren van toedienen van irrigatie water op de productiviteit van Teff aangetoond. Focussen op uitsluitend experimenteel veldonderzoek is een eenzijdige benadering en is nauwelijks voldoende voor het bereiken van de beste oplossingen voor de huidige problemen op het gebied van waterbeheer. Nieuwe richtlijnen voor het gebruik van de gezamenlijke inspanning van experimenteel veldwerk om experimentele veldgegevens te produceren en het gebruik van modellen zijn duidelijk nodig. Het is om deze behoeften en de vereiste toename van Teff productie onder water schaarse omstandigheden dat dit onderzoek haar belangrijkste bijdrage heeft geleverd.

Annex II. About the author

Yenesew Mengiste Yihun was born in Addis Ababa, Ethiopia on 8th of August, 1982. She studied Agricultural Engineering and Mechanization at Debub University, Awassa, Ethiopia, and graduated with BSc.degree in 2004. She was employed by the Ministry of Agriculture and Rural Development in Komobolcha Agricultural Training, Vocation and Educational Training College (KATVET) as Instructor in October 2004. In October 2006, she received scholarship called CIDA-SARIC from Sweden Government to pursue her Master of Science (M.Sc.) study in Haramya University. Then she joined the school of Graduate Studies of Haramya University and graduated in Soil and Water Engineering (Irrigation Engineering) in 2008. After she completed her second degree she moved to Mekelle University, worked as Lecturer and researcher in the college of Dry Land Agriculture and Department of Land Resource Management and Environmental Protection (LaRMEP). In addition to teaching regular courses and undertaking a research, she was also working as higher official of the Mekelle University as Quality Assurance dean of the college of Dry Land Agriculture. Moreover, she has served as principal investigator in Dutch Government sponsored research project - NORAD II on Improving Irrigation Water Management Practices through Deficit Irrigation to attain food security, a case study in Geregera watershed, Atsbi Womberta District, Tigray, Ethiopia.

In July 2009 she got a full scholarship from Netherlands Fellow ship Programme (NUFFIC) for her PhD study in UNESCO-IHE, Institute for water education, Department of Water Science Engineering, Delft, the Netherlands. Her PhD research is on Agricultural Water Productivity Optimization for Irrigated Teff *(Eragrostic Tef)* in a Water Scarce Semi-Arid Region of Ethiopia. She writes a competitive research proposal so that she wins a research fund from International Foundation for Science (IFS).

Journal papers

Yenesew M.Y. and Ketema T., 2009. Yield and water use efficiency of deficit irrigated maize in a semi-arid region of Ethiopia. *Africa Journal of Food, Agriculture, Nutrition and Development*, Volume 9, No. 8, Nairobi, Kenya.

Yihun, Y.M., Haile, A.M., Schultz, B. and Jijo, T.E., 2013. Crop Water Productivity of Irrigated Teff in a Water Stressed Region: *Water resources management*, p. 1-11. DOI: 10.1007/s11269-013-0336-x.

Yihun, Y.M., Haile, A.M., Daniel, B., Schultz, B and Jijo, T.E., 2014. Teff *(Eragrostic Tef)* crop coefficient for effective irrigation planning and management in the semi-arid Central Rift Valley of Ethiopia: Agronomy and Crop science (under review).

Yihun, Y.M., Haile, A.M., Schultz, B and Jijo, T.E., 2014. Application of AquaCrop model under the full and deficit irrigated Teff *(Eragrostic Tef)* in the Central Rift Valley of Ethiopia (under review).

Conference papers

Yihun, Y.M., Schultz, B., Haile, A.M., Jijo, T.E., 2010. Agricultural Water Productivity Optimization in Water Scarce Semi-arid Region of Ethiopia, A preceding in the 61th International Executive Council Meeting (IEC) and the 6th Asian Regional Conference (ARC). of International Conference of Irrigation and Drainage (ICID), Yogyakarta, Indonesia.

Yihun, Y.M., Schultz, B., Haile, A.M., Jijo, T.E., 2011. Optimizing Teff water productivity in water stressed region of Ethiopia. ICID 21st, Tehran, Iran October 2011.

Yihun, Y.M., Schultz, B., Haile, A.M., Jijo, T.E., 2012. Optimizing Agricultural Water Productivity in water scarce region. A poster presented on Water for Food Conference, Blue water and Green Water, May 30,2012 to June 2, 2012,Nebraska, USA.

Manual development

Yihun, Y.M., Bekele, A., Asrat, A., Yesuf, A.,Getachew, A., Demise, A., Wondimu, B., Terefe, B., Yirga, B., Darza, D. et al., 2009. Mainstreaming Gender Awareness in Higher Education. Manual development 2nd Edition at Mekelle University, Mekelle, Ethiopia.

Printed and bound by CPI Group (UK) Ltd, Croydon, CR0 4YY

21/10/2024

01777096-0008